TPNW - Treaty on the Prohibition of Nuclear Weapons

Moving Forward

Published on the 76th Anniversary of the signing of the Nuremberg Charter
8th August 2021

This book complements its 'sister' book, Nuclear Weapons and International Lw - 3rd edition, referred to in this book as **NWIL3.**

Nuclear Weapons and International Law

3rd Edition - major update

Geoffrey Darnton

Foreword and Contributions by Richard Falk and David Krieger
1st edition Foreword and Contributions by Sean MacBride
Contributions by Colin Archer and Nick Grief

TPNW —
Treaty on the Prohibition of Nuclear Weapons

Moving Forward

Geoffrey Darnton

PEACE ANALYTICS

First edition 2021:

978-1-912359-15-8 (Hardback)
978-1-912359-16-5 (Paperback)
978-1-912359-17-2 (eBook)

Peace Analytics (an imprint of Durotriges Press)
Bournemouth, UK

https://tpnw.website/

Contents

What TPNW Does and Does Not Do; TPNW vs NPT vs NWC

The Importance of Domestic Prohibition Measures: the NZ Example

Greater Application of the Nuremberg Principles and ICC

Chapter 7

Case Study: UK and Trident

Chapter 8

Civil Society Manifesto to Move Forward

Appendix A

Treaty on the Prohibition of Nuclear Weapons, 2017

Appendix B

Model Nuclear Weapons, Convention

Appendix C

New Zealand Nuclear Free Zone, Disarmament, and Arms Control Act 1987

Appendix D

North Atlantic Council Statement on the Treaty on the Prohibition of Nuclear Weapons

Appendix E

France, UK, US Statement After UNGA Adoption of TPNW

Appendix F

P5 Statement on TPNW

Preface

The Treaty on the Prohibition of Nuclear Weapons (TPNW), 2017, entered into force on 22nd January 2021, 90 days after the 50th ratification of the Treaty (which happened on 24th October 2020).

There is a need for a summary text that explains the Treaty along with its origins and context. Some commentators comment that they see the Treaty as the 'beginning of the end' of nuclear weapons. This means that further action for nuclear disarmament is inevitable until the goal is reached. Hence, this book is subtitled 'Moving Forward'; the book contains ideas for future activities in support of TPNW. Readers will find a variety of ideas that can be picked up by civil society for future campaigns and activities. TPNW would not have come into existence without a combination of many civil society organizations around the globe, with some international coordination, and skilled diplomats working with governments who wanted to see TPNW as an international treaty. When TPNW came before the UN General Assembly (UNGA) for a vote, in 2017, there was a clear majority of countries in favour of TPNW. The countries that voted against it were the Nuclear Weapons States (NWS) and countries sitting under the umbrella of one of the NWS (umbrella States) , the largest number being members of NATO.

The key plank in the assertions of those countries voting against TPNW was the assertion that TPNW would undermine the Non Proliferation Treaty (NPT) of 1928, and they assert that the NPT is the main instrument providing a global security framework. This book examines that NPT vs TPNW debate. It includes in its appendices key documents issued by members of the UN Security Council (UNSC) and NATO setting out objections to TPNW.

In the course of preparing this book, it was not possible to find any coherent argument or legal argument about how TPNW could undermine the NPT. It is concluded that TPNW complements and reinforces the NPT - in fact, it shows many countries doing their best to meet their legal obligations arising from NPT. Therefore it is necessary to conclude that the NWS and umbrella States prefer a regime where they have a cynical disregard for legal obligations of the NPT and prefer to maintain the NPT status quo. The cynical disregard for legal obligations of the NPT is itself the greatest threat to the NPT framework because the reckless behaviour for more than 50 years now provides so many countries with a reason to withdraw from the NPT..

Another cynical argument of the NWS and umbrella States is that nuclear weapons *per se* are not prohibited by current treaty law. They cynically avoid some treaty law such as the UN Charter and the Geneva Conventions, along with a great deal of customary international law as examined by the International Curt of Justice (ICJ) in their Advisory Opinion (ICJ AO) of 1996. There are no circumstances in which nuclear weapons could be used lawfully. This book picks up the idea of a Nuclear Weapons Convention (NWC) that could sit side by side with other conventions covering biological and chemical weapons.

When putting this book together, there was a dilemma. The sister book, Nuclear Weapons and International Law - 3rd edition (NWIL3), was published on 8th August 2020. That contains a chapter about the proposed TPNW. Since August 2020, TPNW entered into force in January 2021. therefore the choice was to update NWIL3 to NWIL4. NWIL3 is already a substantial book. Therefore the decision was taken to do a separate smaller book as a summary of relevant topics. This keeps the cost down for the TPNW material. However, you will find NWIL3 a very suitable supplement to this book, for a lot more further information.

This book is suitable for lawyers (academic and practitioner) civil society activists, and academics.

Geoffrey Darnton
Bournemouth,
8th August, 2021
(76th Anniversary of the signing of the Nuremberg Charter)

Acknowledgements

Many people have contributed to this book. Some indirectly via the references used in the book or being part of the teams producing documents in the Appendices, and others by comments and suggestions on drafts of parts of the book or by participating in email discussions and threads related to progress of the book.

To mention but a few: **Colin Archer, Kate Dewes**, Richard Falk, **Robert Forsyth, Robert Green**, Nick Grief, John Hallam, Kate Hudson, Mike Kiely, Sue Miller, Nick Ritchie, Matt Robson, Keith Suter, Aaron Tovish, Rob van Riet, Alyn Ware, Angie Zelter (those in bold helped with commenting and/or proofreading chapters). Apologies to anyone not mentioned by name - there are many.

Full editorial responsibility for the contents of the book has remained with the Author in all chapters. The book cover design was provided by Akhtanbed of India through DesignCrowd.

Acronyms, and Abbreviations

AWE ....Atomic Weapons Establishment (in UK)
BWC ....Biological Weapons Convention
CND ....Campaign for Nuclear Disarmament
CWC   Chemical Weapons Convention
DAC..............Direct Action Committee
IALANA.......International Association of Lawyers Against Nuclear Weapons
ICAN ....International Physicians for the Prevention of Nuclear War
ICC ....International Criminal Court
ICJ................International Court of Justice
ICJ AO.........International Court of Justice Advisory Opinion 1996
ILC...............International Law Commission
IMT ....International Military Tribunal(Nuremberg or Tokyo)
IPB................International Peace Bureau
IPPNW.........International Physicians for the Prevention of Nuclear War
LNWT.........London Nuclear Warfare Tribunal
MDA.............Mutual Defence Agreement (UK and USA)
MAD.............Mutually Assured Destruction
MID ....Massive Interventions of Democracy
NAPF...........Nuclear Age Peace Foundation
NATO...........North Atlantic Treaty Organization
NPT ....Non-Proliferation Treaty 1968
NTC..............Nuclear Terrorism Convention, 2005
NWC ....Nuclear Weapons Convention
NWIL3 ....Nuclear Weapons and International Law 3rd Edition
NWS.............Nuclear Weapon States
NZ ....New Zealand
OEWG ....Open-Ended Working Group
P3 ....France, UK, USA - members of UNSC
P5 ....Five Permanent embers of the UN Security Council
PNND...........Parliamentarians for Nuclear Non-Proliferation and Disarmament
RCW..............Reaching Critical Will
RMI ....Republic of the Marshall Islands

x

SDI.................Strategic Defence Initiative
TPNW..........Treaty om the   Prohibition of Nuclear Weapons 2017
UNGA ....United Nations General Assembly
UNSC ....United Nations Security Council
WCP..............World Court Project
WNA..............World Nuclear Association
UK.................United Kingdom
USA..............United States of America
USSR.............Union of Soviet Socialist Republics
VCLT ....Vienna Convention on the Law of Treaties
WILPF..........Women's International League for Peace and Freedom
WWII.............2nd World War (1939-45)
NWFZ...........Nuclear Weapon FreeZone

Chapter 1

Introduction

1.1 Reading This Book

This book uses summaries of, and makes frequent reference to it's 'sister' book, *Nuclear Weapons and International Law: 3rd edition*, by Darnton et al. (2020). Throughout this books the sister book is referred to as NWIL3.

This book is stand-alone. However, you may want more background for some of the topics, in which case use NWIL3 or the many relevant sources that are readily available. Many have pointers in the References.

1.2 What is in This Book?

Chapter 2 discusses key milestones on the long road from the birth of the bomb to TPNW. Depending on your starting point, that road is about 80 years. Serious research and recognition of the possibility of a 'super' (nuclear) bomb started around the late 1930s/early 1940s.

The beginnings of the United Nation were in 1941 with the Inter-Allied Declaration and Atlantic Charter followed by the Declaration of the United Nations - 26 countries - on 1st January 1942.

There were several key events in 1945: the UN Charter was signed in June 1945 and entered into force on 24th October 1945; Hiroshima was atomic bombed on 6th August 1; the Nuremberg Charter signed on 8th August; Nagasaki atomic bombed on 9th August.

Following 1945 there was proliferation of nuclear weapons to all the UNSC permanent members (P5). Due to increasing concern about nuclear weapons proliferation, and the desire of the P5 to maintain a 'closed-shop' of NWS, the Non-Proliferation Treaty was signed in 1968. Civil society anti-nuclear organizations around the world started activities and protests; the LNWT was held in London January 1985; the WCP ran from 1991 to 1995; the ICJ delivered its opinion; global dissatisfaction and frustration at the NWS lack of action and good faith by the NWS produced a build up of 'steam; between 2013 and 2017, TPNW was negotiated and entered into force in January 2021.

2

During the whole period 1940-2021, nuclear rectors, which are the source of core materials for nuclear weapons, has escaped scrutiny under the guise of 'peaceful' uses of nuclear energy.

The chapter also discusses problems with the NPT and the ICJ AO.

TPNW can be viewed more as a codification of existing international law, rather than creating new law. Nuclear Weapons were already illegal before TPNW. TOne novelty in TPNW is the requirement for States Parties to implement treaty obligations in domestic law. This position is explored in Chapter 3, and Chapter 5 discusses the issue of domestic law further, using New Zealand as an example. Chapter 3 includes the rationale for considering 'nuclear deterrence' to be unlawful.

Chapter 4 sets the scene to move forward from TPNW by looking more closely at what TPNW does and does not do. It looks in more detail at the debate about TPNW vs NPT, and uses proposals for a Nuclear Weapons Convention (NWC) as a framework for a future international treaty.

So far, most narrative is focussed on the behaviour and activities of nation states. Extremists like to talk about sovereign states, but of course, willingly entering into international treaties is agreeing to sovereignty compromises. Of course, talking about states doing or not doing something is fundamentally anthropomorphic; states cannot do something - only individuals can. This anthropomorphic dilemma can only be resolved by the legal device of considering a sovereign state as a separate legal entity. That is normally not problematic, but is so in some cases (for example, Palestine). The international criminal liability of individuals was established in international law back in 1945 by the Nuremberg Charter, and subsequently reinforced by the Rome Statute that set up the International Criminal Court (ICC). Chapter 6 discusses this issue of individual liability with respect to nuclear weapons. As nuclear weapons are inherently illegal and pose an existential threat to humanity and the environment, all individuals responsible for taking decisions to develop, finance, manufacture, acquire, deploy, maintain, threaten to use, or use etc., assume personal individual liability under international law for their actions. One way a country can work to protect all its citizens is to become a state party to TPNW and implement appropriate domestic legislation.

In March 2021, the UK Prime Minister made a shock announcement the the UK Government had carried out an Integrated Review of defence policy. That Review included an announcement that the UK would increase (another anthropomorphism - in reality it is particular individuals who carried out the review and took the reported decisions - and who will be implementing them).

As the UK has been involved intimately with the USA since the 'birth of the bomb' (Hiroshima and Nagasaki would probably not have been possible by 1945 without the help of British scientists working on the Manhattan Project), Chapter 7 presents a case study of UK nuclear weapons policy from the UK decision to develop its own 'independent' nuclear weapon system to the recent UK Government Integrated Review.

The subtitle of this book is 'Moving Forward', therefore the final chapter (8) collates issues from throughout the book for ways to move forward from TPNW. If there is any sense in which TPNW can be seen as the 'beginning of the end' for nuclear weapons, some of the proposed ideas are inevitable; some are more radical. Some of the suggestions will require diplomatic effort; some will require civil society engagement, and, of course, some of the needed diplomatic work will need a lot of prompting and engagement of coordinated civil society action.

A book like this needs supporting material to understand its underpinnings and relationships with the works of others. That is provided in a variety of ways. The book is written in an academic style with references and in-text captions pointing to related references. The referencing style adopted is author-date with references at the end collated by family name of first author; some of the references are 'invented' to simplify referencing, the fundamental principle being that an in-text citation can be found easily in References for more detailed information. Although the style is more a politico- or socio-legal text, than a purely legal text, it avoids almost completely, the footnote referencing style of legal works. This makes it much easier to obtain an holistic view of the totality of references used. it has References, not Bibliography, as each reference must be used at least once in the body of the book text.

There is the associated 'sister' book NWIL3 (Darnton et al, 2020) which has more detail of many of the themes introduced in this book

Essential supporting material is included in the Appendices to this book.

Appendix A is the full text of the TPNW - essential reading to underpin this book.

One key step needed by the nations of the world, is to negotiate a Nuclear Weapons Convention, to sit with the Conventions covering biological and chemical weapons NWC is discussed in the book and compared to TPNW. The latest draft NWC is shown in Appendix B. That draft will require amendment by negotiation.

While TPNW was being negotiated, all the NWS failed to participate in the negotiations, although they voted in the UNGA when the negotiated Treaty came up for its final vote.

As can be imagined, the NWS are not happy about TPNW and there was a lot of background lobbying by NWS of non-NWS to try to persuade them not to sign or ratify.

New Zealand set up its own nuclear weapons free zone and implemented a series of related treaties in domestic law, so the text of the relevant 1987 New Zealand Act is given in Appendix C (those related treaties snipped from the Appendix).

The principle argument of NWS and umbrella States was that the TPNW acts to undermine the NPT. In order to understand the arguments used by some NWS, Appendix D gives the text of the NATO position via the North Atlantic Council, Appendix E gives the text of France, UK, and USA (P3) and Appendix F gives the text of the P5. The book shows that these NWS statements give no rational basis for their assertion about TPNW undermining the NPT, and the book shows that TPNW does part of what the NWS and umbrella states should have achieved long ago. The NWS want to cling on to the NPT by completely avoiding their cl;ear legal obligations. There is absolutely nothing in the NPT that prevents a state from joining the TPNW; indeed, Article VII gives the right to states to form NWFZs:

"Article VII
Nothing in this Treaty affects the right of any group of States to conclude regional treaties in order to assure the total absence of nuclear weapons in their respective territories."

Every time a country joins the TPNW it is acting completely in accordance with parts of the NPT; there is no undermining; there is simply partial implementation of the duty to create a nuclear weapon free world. It is the NWS and umbrella states who are undermining the NPT by their chronic failure (>50 years!) to even attempt in good faith to meet their legal obligations under the NPT. The NWS and umbrella states are in material breach of the NPT: the TPNW provides a mechanism by which states can demonstrate their good faith in working towards a nuclear weapon free world doing what they can as individual states working together with others. Not only are the NWS among the world's biggest bullies and terrorists, they are also among the world's greatest contributors to climate change.

The TPNW Journey from the Birth of the Bomb

2.1 Introduction

TPNW was born following decades of frustration expresses by large segments of civil society, a majority of UN member states, many non-governmental organizations, and the public conscience at as a whole. This was frustration at a world order that enables the nuclear weapons states to pose an existential threat to humanity and the planet by their weapons of mass destruction, and their abject failure to negotiate in good faith for binding multilateral nuclear disarmament, notwithstanding their clear legal duty to do so.

As explained in the next chapter, nuclear weapons were already illegal under international law before TPNW put an important stake in the ground. Humanity must move forward and force the leaders of the nuclear weapon states to ensure that they negotiate in good faith for total nuclear disarmament.

For all the anti-terrorist bluster from leaders of the nuclear weapons states (NWS), the reality is stark - the NWS are the largest terrorist organizations on the planet - in a purely technical sense by the meaning of 'terrorism', not in a pejorative sense. The leaders of the NWS are terrorist leaders; they threaten others with the use of nuclear weapons. The United Nations is neutered because it is hijacked by these terrorists who hold veto powers.

Since the first use of nuclear weapons, demand for their total elimination has been growing. The early days saw some signs of optimism that then dissipated as nuclear weapons states failed in their legal duty to negotiate full multilateral disarmament. There has been an increasing frustration by civil society and a majority of governments with proliferation and lack of progress. This was made worse by the absense of any participation (except behind the scenes lobbying and sometimes bullying), by the NWS as progress towards TPNW built up. Even though the NWS would, understandably, not want to see TPNW enter into force, their failure to participate at all remains shameful.

TPNW is a major consolidation of earlier concerns and includes a whole range of important issues related to nuclear weapons. Many see

6

TPNW as the beginning of the end of nuclear weapons, even though that end may still be many years in the future.

For the purposes of this Chapter, the TPNW journey is set out very briefly in a series of phases, including brief discussions of problems with the Non-Proliferation Treaty (NPT), 1968, and judgments of the International Court of Justice (ICJ). Much more detail about most of these phases can be found in the sister book NWIL3 and the other works referred to. There is considerable timing overlap between the phases presented here. The phases are:

o   **Birth of the Bomb**
o   **Hiroshima and Nagasaki**
o   **Birth of the United Nations**
o   **Birth of More Nuclear Weapons States**
o   **General Assembly Deliberations**
o   **Civil Society Mobilization**
o   **Non-Proliferation Treaty**
o   **Nuclear Weapons Free Zones**
o   **Proliferation and Unilateral Disarmament**
o   **London Nuclear Warfare Tribunal**
o   **World Court Project**
o   **International Court of Justice**
o   **Renewed Mobilization of Civil Society**
o   **Birth of TPNW**
o   **Moving Forward from TPNW**

2.2 Birth of the Bomb

Uranium was discovered towards the end of the 18th century. In the 19th and early 20th centuries a number of scientists was involved in a series of experiments, research, and developments that proved central to the development of nuclear weapons (WNA, 2020).

There was a dramatic increase in research and scientific investigation in the 1930s (Rhodes, 1996; Gale, 2001) in a number of countries such as Germany, (Porter, 2010). Soviet Union (Charles Rivers Editors, 2019), UK (Clark, 1961; Szasz, 1992), and USA (Kelly, 2007). Other countries proved to be pivotal; Norway had the world's largest heavy water production plant used in fertilizer production, but in World War II Germany understood the significance of heavy water for the potential it offered in its nuclear bomb project - which the British also understood; hence the 1942 Allied raid on the Norwegian facility at Telemark.

The developments in World War II meant there was already a nuclear arms race between Germany and USA-Britain. The Soviet Union had suspended its work due to the German invasion. This early nuclear weapons work was clouded in considerable secrecy and espionage.

The largest WW II project was the Manhattan project which was a joint USA-British project. For a discussion of Manhattan see works such as Kelly (2009), and for the role Britain played, see works such as Gallagher (2002).

The first nuclear bomb test was at Alamagordo in New Mexico on 16 July 1945 using a plutonium bomb.

2.3 Hiroshima and Nagasaki

The deep secrecy during WWII by all the powers involved in trying to build an atomic bomb meant that there was very little knowledge of nuclear weapons or their effects, and no possibility to assess their legality under international law other than by those involved in the design and development,

The atomic bombing of Hiroshima and Nagasaki thrust the effects of nuclear weapons into the public domain forcefully. No more absolute secrecy. The debates about the effects, rights, and wrongs of nuclear weapons began almost immediately because they were such terrible weapons.

The bombs dropped on Hiroshima and Nagasaki were the principal products of the Manhattan Project. That Project had the USA as the major partner and Britain as a minor partner (see Szasz 1992 for information about British scientists involved) In the later stages of the Manhattan Project, the USA and Britain were also working on the development of the Nuremberg Charter, starting from before the Moscow Declaration in 1943 (see Moscow, 1943). As the USA and Britain were involved in the emerging Charter, they were fully aware of the key principles of international law which formed the basis of it. See Duling (2019) on the order to drop the bomb.

It was with deep irony and cynicism that the USA dropped the atomic bomb on Hiroshima on 6th August 1945, Britain and the USA signed the Nuremberg Charter on 8th August 1945, and the USA dropped another atomic bomb on 9th August 1945. Dropping those bombs on Hiroshima and Nagasaki were grave violations of international law as it stood at the time. Not only did the USA know it, the USA had signed the Nuremberg Charter in between the bombings. Nobody in the USA or Britain has ever been held accountable under international criminal law for their individual participation in the design, development, and use of those atomic bombs, nor have the governments. The Tribunals held in accordance with the Nuremberg Charter were, in Justice Pal's terms, "victors' justice" (Pal, 1948). However, although Pal was concerned at the absence of the USA and Britain, he appears not to have given due consideration either to the Japanese atrocities of WWII.

There is still secrecy and lack of transparency about the decisions taken to drop the bombs. The USA decision makers clearly didn't care about the law breaking of which they were fully aware. The later arguments of military necessity simply don't wash as Japan was on the point of surrender (for an example of debates around this, see Alperovitz, 1995; Villa and Bonnett, 1996). The two bombs used different technology; one was uranium, the other was plutonium. It is widely assumed that these were further experimental explosions. At that stage, the USA decision makers didn't know about the radiation effects that would follow the bombings. There were also important geo-political factors influencing the situation. The Soviet Union had just entered the war against Japan. The USA had already seen the results of the Soviet Union takeover of European countries and did not want a Soviet takeover of Japan; Japan preferred takeover by the USA to takeover by the Soviet Union.

There would still be merit in a public investigation of the bombings of Hiroshima and Nagasaki and some of the other allied military operations during WWII. Maybe another civil society tribunal?

There are many sources of information about the terrible effects of the bombings of Hiroshima and Nagasaki. Some of those effects continue to this day.

The lack of Japanese respect of international humanitarian law by both Japanese and German military forces and governments is well documented; the lack of respect for international law by allied military forces and governments still awaits full investigation.

2.4 Birth of the United Nations

While work was under way building the first nuclear bombs and developing the Nuremberg Charter to make individuals criminally liable for breaches of international criminal law, the final touches were being put to the United Nations. Charter. Work had started in 1941, it was signed on 26th June 1945, and came into force on 24th October 1945.

The UN Charter is an international treaty with almost all countries of the world being states parties.

It is worth recalling the principal goals of the UN as set out in the Preamble to the Charter:

"WE THE PEOPLES OF THE UNITED NATIONS DETERMINED

to save succeeding generations from the scourge of war, which twice in our lifetime has brought untold sorrow to mankind, and
to reaffirm faith in fundamental human rights, in the dignity and worth of the human person, in the equal rights of men and women and of nations large and small, and
to establish conditions under which justice and respect for the obligations arising from treaties and other sources of international law can be maintained, and
to promote social progress and better standards of life in larger freedom,

AND FOR THESE ENDS

to practice tolerance and live together in peace with one another as good neighbours, and
to unite our strength to maintain international peace and security, and
to ensure, by the acceptance of principles and the institution of methods, that armed force shall not be used, save in the common interest, and
to employ international machinery for the promotion of the economic and social advancement of all peoples,

HAVE RESOLVED TO COMBINE OUR EFFORTS TO ACCOMPLISH THESE AIMS

Accordingly, our respective Governments, through representatives assembled in the city of San Francisco, who have exhibited their full powers found to be in good and due form, have agreed to the present Charter of the United Nations and do hereby establish an international organization to be known as the United Nations."

Legal purists may argue that the Preamble is not a binding part of the treaty; it merely summarizes key points included in the actual treaty.

On the face of it, this is a wonderful document for generating a widespread hope among the bulk of humanity. Freedom from war, adherence to international law, greater freedom, reductions in poverty, and so forth.

The reality since 1945 has been continuous war, breaches of international law by many countries and their leaders, less freedom for many, and other problems suffered as a consequence of very poor leadership in many countries.

The UN Charter has been honoured more in the breach than its fulfilment. It seems as though the five permanent members of the UN Security Council are the worst terrorist organizations on the planet!

The frequent serious failures of many states party to the UN Charter mark the beginning of the frustrations of civil society and many compliant countries contributed to the start of the long journey to TPNW.

On 24th January 1946, the United Nations General Assembly (UNGA) adopted by consensus its very first resolution Resolution 1 (I), which established a commission of the UN Security Council to ensure 'the elimination from national armaments of atomic weapons and all other major weapons adaptable to mass destruction.'

This resolution was adopted unanimously, including by the only known nuclear weapons state at the time - the USA.

24th January 2021 was the 75th anniversary of the passing of that UNGA Resolution 1(1). Not only does the USA still have nuclear weapons, there has been considerable proliferation, all in breach of obligations under the UN Charter, international law, and many further UNGA Resolutions. The NPT put the brakes on, but as noted below, did not stop it completely.

This continuing situation of multiple states compounding their breaches of international law is further background to frustrations leading to TPNW.

2.5 Birth of More Nuclear Weapons States

At the time of the entry into force of the UN Charter and UNGA Resolution 1(1), there was only one known nuclear weapons state - USA.

Use whichever metaphor you prefer; the bombs dropped on Hiroshima and Nagasaki let the genie out of the bottle and opened Pandora's box.

There was a scramble by other countries to have the capability to develop, build, and deliver nuclear weapons. Britain had been part of the Manhattan Project. The Soviet Union had put its own nuclear research on hold after being invaded by Germany, but by 1945 knew about atomic development in Germany and espionage activity had yielded information about the Manhattan Project. The USA did not help the situation when some hot head military and politicians drew up a list of targets in the Soviet Union for possible atomic attack. The Soviet Union carried out its first atomic test on 29th August 1948 at

Semipalatinsk in Kazakhstan.

Britain started its own independent atomic weapons program in 1947 and conducted its first test at the Montebello Islands in Western Australia on 3rd October 1952.

France conducted its first nuclear test on 13th February 1960 at the Reganne Oasis, in the Algerian Sahara.

China conducted its first nuclear test on 16th October 1964 at Lop Nor located between the Taklamakan and Kumtag deserts in the southeastern part of the Xinjiang.

Therefore, by 1964 the UN Security Council's five permanent members, each holding a veto, had become nuclear weapons states content to breach their international legal obligations and possess nuclear weapons. This rendered the UN incapable of controlling nuclear weapons. This also meant that the world order and the UN had been hijacked by the biggest terrorist states in world history.

This situation added to the frustrations of civil society and a majority of the world's nations - yet another reason giving impetus to the TPNW journey.

Since their first atomic tests, all these initial five nuclear weapons states and permanent members of the UN Security Council have carried out many additional tests, the largest number by the USA at over 1,000. The tests have left a terrible legacy for people in the testing areas, particularly South Pacific Islands and the Semipalatinsk region of Kazakhstan.

2.6 General Assembly Deliberations

Since the start of the UN General Assembly with its 1st session in 1946, there have been repeated resolutions calling for nuclear disarmament.

24th January 2021 saw the 75th anniversary of UNGA Resolution 1(1). The vast majority of the world's countries want global nuclear disarmament. The nuclear weapons states fail to act as responsible global players and remain indifferent to the frequent UNGA calls for disarmament.

This continued failure for more that three quarters of a century by nuclear weapons states combined with further nuclear weapons proliferation and appallingly low respect by nuclear weapons states for nuclear free zones, , has resulted in a significant number of UNGA countries being all the more determined to eliminate all nuclear weapons

UNGA had no difficulty finding the votes to set up the negotiations for, and pass, the negotiated TPNW notwithstanding the impossibility of securing UN Security Council support (because of the vetoes held by nuclear weapons states) or countries sitting under the nuclear umbrella of one of the nuclear superpowers.

2.7 Civil Society Mobilization

Civil society has been mobilized against war for a long time before the age of nuclear weapons. Two of the oldest international organizations are the International Peace Bureau (IPB) and the Women's International League for Peace and Freedom (WILPF) - see Chapter 10 about IPB in NWIL3 (Archer, 2020).

As well as international organizations picking up the challenge of nuclear weapons, there are many organizations at a national and regional level which also picked up the challenges of nuclear weapons. A very good source of information about peace organizations generally, and organizations around the world focussed on nuclear weapons is the Housmans World Peace Directory (see Housmans, 2021) that lists over 1,000 peace organizations worldwide. There is no centralized or controlling organization for these organizations. They show the considerable variety of concerns about peace and nuclear weapons.

One example of a national organization is in the UK: the Campaign for Nuclear Disarmament (CND). 1958 was a pivotal year; the Direct Action Committee (DAC) morphed into the Campaign for Nuclear Disarmament. The nnual Aldermaston March — a gruelling 4-day march by thousands of marchers between London and Aldermaston, the nuclear warheads manufacturing site - was held over several years DAC and CND started a few years after the first British nuclear weapon test in 1952. For more detailed information about CND see sources such as Hudson (2019).

From time to time, multiple organizations come together under an umbrella organization for a particular focus. Examples of these are: Abolition 2000, International Peace Bureau (IPB); International Association of Lawyers Against Nuclear Weapons (IALANA); International Physicians for the Prevention of Nuclear War (IPPNW); Nuclear Age Peace Foundation (NAPF); Parliamentarians for Nuclear Non-Proliferation and Disarmament (PNND), World Court Project (see below); and many others.

As far as the journey to TPNW is concerned, many organizations operated at a national level in debates with governments about voting in the UNGA stages leading up to TPNW. For example, under the auspices of the Women's International League for Peace and Freedom is Reaching Critical Will (RCW) which was instrumental in organizing civil society participation in the UN working groups and negotiations that led to TPNW. The International Campaign to Abolish Nuclear Weapons (ICAN) played the leading role as an umbrella organization pressing for TPNW, for which it was awarded the Nobel Peace Prize in 2017.

2.8 Non-Proliferation Treaty

By the 1960s most countries in the world agreed there was a need to prevent the further proliferation of nuclear weapons, and for the negotiation for complete multilateral nuclear disarmament.

The result of these deliberations and negotiations was the Treaty on the Non-Proliferation of Nuclear Weapons (NPT) in 1968

All the nuclear weapons states at the time signed and ratified the Treaty, thus paying at least lip service to the aims and promises of the Treaty. See Popp et al. — negotiating the Treaty. Non-nuclear weapons states were seduced into the treaty by the promises of Article VI:

> *Article VI*
> Each of the Parties to the Treaty undertakes to pursue negotiations in good faith on effective measures relating to cessation of the nuclear arms race at an early date and to nuclear disarmament, and on a treaty on general and complete disarmament under strict and effective international control.

Note a couple of important points in this: "in good faith", and, "at an early date" (for further discussion about these points, see Farebrother, 2012). There has been total failure by all the nuclear weapons states to meet their obligations to negotiate in good faith and at an early date.

The NPT has not been successful in suppressing all proliferation, but has mainly kept the lid on it.

North Korea was a signatory to the NPT but withdrew in 2003. India, Israel, and Pakistan did not sign the NPT, presumably because by 1968 they were aware of their own, and their neighbours' nuclear weapons ambitions. This total failure by the NWS to meet their NPT obligations has been an increasing source of frustration among non-NWS and provided the main impetus for the journey to TPNW.

14
2.9 Nuclear Weapon Free Zones

Further frustration in many countries given their wish to see a world free from nuclear weapons spurred many countries to establish nuclear weapon free zones (NWFZs).

Setting up NWFZs is completely consistent with article VII of the NPT.

The first NWFZ was established just before (1967) the NPT. In chronological order, the NWFZs are: Antarctic Treaty, 1959; Treaty on Principles Governing the Activities of States in the Exploration and Use of Outer Space, including the Moon and Other Celestial Bodies (Outer Space Treaty) 1967; Treaty for the Prohibition of Nuclear Weapons in Latin America (Tlatelolco Treaty) 1967; Treaty on the Prohibition of the Emplacement of Nuclear Weapons and Other Weapons of Mass Destruction on the Sea-Bed and the Ocean Floor and in the Subsoil Thereof (Seabed Treaty) 1971; Agreement Governing the Activities of States on the Moon and Other Celestial Bodies (Moon Treaty), 1979; South Pacific Nuclear Free Zone Treaty (Rarotonga Treaty) 1985; New Zealand Nuclear Free Zone, Disarmament, and Arms Control Act 1987; Southeast Asian Nuclear-Weapon-Free Zone Treaty (Bangkok Treaty) 1995; African Nuclear-Weapon-Free Zone Treaty (Pelindaba Treaty) 1995; Mongolia, 2000; Treaty on a Nuclear-Weapon-Free Zone in Central Asia (Semipalatinsk Treaty) 2006;. Since 2017, States Parties to TPNW establish their own individual NWFZ. (See also NWI3, Chapter 16.)

2.10 Proliferation and Unilateral Disarmament

Nuclear weapons proliferation started well before the NPT. Israel started work on establishing its nuclear energy program around the time of the formation of the Israeli state in 1948. In those days 'peaceful' nuclear energy generation was just a cover for the production of fuel needed for nuclear weapons. At the beginning, producing electricity from nuclear reactors was simply a by-product of producing nuclear weapons material. The Israeli nuclear programme received help from France and the USA. Heavy water and uranium production were started in the early phases. Israel always ran its nuclear programme in great secrecy and even today operates a policy of neither confirming nor denying its nuclear weapons. Other states are complicit in this public secrecy, in part to cover their ownrole in assisting Israel's nuclear weapons programme.

Despite the NPT, nuclear weapons spread to more countries. Doubtless some of this happened because of some of the nuclear weapons states who were parties to the NPT being willing to breach, in secret, their own responsibilities under the NPT.

The first Israeli nuclear tests were held in 1960-63, with one of them possibly in collaboration with France. See sources such as Jabber (1971) and, Gilling and McKnight (1991).

India's first nuclear weapon test was on 18th May 1974 at the Pokhran Test Range, in Rajasthan.

Pakistan conducted its first nuclear weapon tests in 28th May 1998 at Ras Koh Hills in the Chagai District of Baluchistan.

Israel, India, and Pakistan were not parties to the NPT, but some countries that helped them were.

North Korea withdrew (2003) from the NPT and carried out its first nuclear weapon test on 9th October 2006 at its Punggye-ri nuclear test site. During the Trump USA Presidency, there were talks between the USA and North Korea but would not have made much progress while there remained a fundamental misunderstanding between the parties about what each was talking about; the USA was talking about North Korean nuclear disarmament while North Korea was talking about nuclear disarmament of the whole Korean peninsula. There is an intriguing discussion about how a secret program by Japan to build its own nuclear weapon was a motivator for North Korea to start its own nuclear programme (Wilcox, 2019).

There have been cases of unilateral nuclear disarmament. The break up of the Soviet Union starting around 1989 could have led to a sudden proliferation of NWS because of the nuclear weapons in several; former Soviet Republics. What is not clear are the weapons control systems, i.e. whether those countries could ever have used the nuclear weapons on their territory. Whatever the situation, three former Soviet Republics decided they didn't want nuclear weapons: Belarus, Kazakhstan, and Ukraine. Kazakhstan went on to sign up to a NWFZ. The missiles were transferred to the Russian Federation, and some nuclear materials to the USA.

Someone should make a movie about the story of South Africa's nuclear weapons development and subsequent unilateral nuclear disarmament. The narrative involves France, USA, Israel, Pakistan, Soviet Union, Cuba, and some others. The whole story involves not only the matter of developing and dismantling nuclear weapons but also geo-political factors involving other African countries, with the main reason to prevent nuclear weapons falling into ethnic African control. The net effect was that South Africa developed nuclear weapons, went through unilateral disarmament in 1989, and signed the NPT in 1991. Subsequently, South Africa supported the development of, and ratified, TPNW. See sources such as Venter (2008) and the very many sources shown in Wikipedia contributors (2021, February 15).

So, two of these countries that have disarmed unilaterally have been involved in tests, have signed and ratified TPNW and both have a NWFZ with the shortened treaty name using a location on their territories (Kazakhstan and South Africa).

2.11 London Nuclear Warfare Tribunal

The London Nuclear Warfare Tribunal (LNWT) was planned during 1984 following an initiative of the Ecology Party (now known as the Green Party), and was conducted in London during January 1985.

It was run with an eminent and experienced set of lawyers and practitioners who conducted a detailed investigation of the legal issues surrounding nuclear weapons. It served as a launchpad for several initiatives that eventually led to TPNW. See more information in Darnton (1989) which is the 1st edition of NWIL3, the sister book to this one.

2.12 World Court Project

Many people for many years had the hope of a formal legal decision outlawing nuclear weapons.

Seán MacBride (IPB President and Nobel Laureate) was an early advocate of a reference to the ICJ as part of his determination to strive for nuclear disarmament (Dewes, 1998). MacBride was Chairman of the 1985 London Nuclear Warfare Tribunal. As early as 1974, MacBride called for a convention to outlaw the use of nuclear weapons in his Nobel Peace Prize address (MacBride, 1974).

The fifth Recommendation the Judgment and Recommendations of the London Nuclear Warfare Tribunal, is:

"5. The initiation of an effort to obtain an Advisory Opinion of the International Court of Justice on the status of nuclear weapons, strategic doctrines, and war plans;"

The World Court Project was launched internationally in Geneva in May 1992, and the ICJ Advisory Opinion delivered in 1996: a remarkably short time scale for this project to have succeeded. Further information about the birth of the WCP is available in NWIL3 Chapters 9 & 10.

To understand the significance of the WCP, there were more than 700 organizations in more than 80 countries involved, and nearly 4 million individuals who registered their concerns about the continued possession and possible use of nuclear weapons.

Diplomatic efforts resulted in WHO and UNGA references to the ICJ for an advisory opinion about the legality of nuclear weapons.

The matter was referred to the International Court of Justice (ICJ) by the UN General Assembly.

2.13 International Court of Justice

The ICJ has considered two cases of considerable significance to those concerned about nuclear weapons: (1) the WHO/UNGA requests for an Advisory Opinion about the legality of nuclear weapons; and, (2) the 2014 case of the Marshall Islands against nuclear weapons states concerning their failure to meet their obligation under Article 6 of the NPT to negotiate in good faith for multilateral nuclear disarmament.

ICJ decided it would not consider the request by WHA/WHO for an Advisory Opinion about nuclear weapons as it considered that WHA/WHO did not have the jurisdiction to hear the case on the grounds that the subject matter was outside the scope the work of WHA/WHO.

At the end of 1994, the General Assembly of the United Nations decided to ask the International Court of Justice for an advisory opinion:

'Is the threat or use of nuclear weapons in any circumstance permitted under international law?'

The full ICJ Advisory Opinion can be found in NWIL3 Chapter 11, so here will be listed the final summary of key points in the Opinion.

THE COURT,

(1) By thirteen votes to one,

Decides to comply with the request for an advisory opinion;

IN FAVOUR: *President* Bedjaoui; *Vice-President* Schwebel; *Judges* Guillaume, Shahabuddeen, Weeramantry, Ranjeva, Herczegh, Shi, Fleischhauer, Koroma, Vereshchetin, Ferrari Bravo, Higgins;
AGAINST: *Judge* Oda;

(2) *Replies* in the following manner to the question put by the General Assembly:
A. Unanimously,

There is in neither customary nor conventional international law any specific authorization of the threat or use of nuclear weapons;

B. By eleven votes to three,

There is in neither customary nor conventional international law any comprehensive and universal prohibition of the threat or use of nuclear weapons as such;

IN FAVOUR : *President* Bedjaoui; *Vice-President* Schwebel ; *Judges* Oda, Guillaume, Ranjeva, Herczegh, Shi, Fleischhauer, Vereshchetin, Ferrari Bravo, Higgins;
AGAINST : *Judges* Shahabuddeen, Weeramantry, Koroma;

C. Unanimously,

A threat or use of force by means of nuclear weapons that is contrary to Article 2, paragraph 4, of the United Nations Charter and that fails to meet all the requirements of Article 51, is unlawful;

D. Unanimously,

A threat or use of nuclear weapons should also be compatible with the requirements of the international law applicable in armed conflict, particularly those of the principles and rules of international humanitarian law, as well as with specific obligations under treaties and other undertakings which expressly deal with nuclear weapons;

E. By seven votes to seven, by the President's casting vote,

It follows from the above-mentioned requirements that the threat or use of nuclear weapons would generally be contrary to the rules of international law applicable in armed conflict, and in particular the principles and rules of humanitarian law;
However, in view of the current state of international law, and of the elements of fact at its disposal, the Court cannot conclude definitively whether the threat or use of nuclear weapons would be lawful or unlawful in an extreme circumstance of self-defence, in which the very survival of a State would be at stake;

IN FAVOUR : *President* Bedjaoui; *Judges* Ranjeva, Herczegh, Shi, Fleischhauer, Vereshchetin, Ferrari Bravo;
AGAINST : *Vice-President* Schwebel; *Judges* Oda, Guillaume, Shahabuddeen, Weeramantry, Koroma, Higgins;

F. Unanimously,

There exists an obligation to pursue in good faith and bring to a conclusion negotiations leading to nuclear disarmament in all its aspects under strict and effective international control.

Done in English and in French, the English text being authoritative, at the Peace Palace, The Hague, this eighth day of July, one thousand nine hundred and ninety-six, in two copies, one of which will be placed in the archives of the Court and the other transmitted to the Secretary-General of the United Nations.

The decisions are very clear relating to the threat or use of nuclear weapons, and the requirements of the NPT. The only problem is in 2E where there is a statement that the Court cannot make up its mind in the case of "in an extreme circumstance of self-defence, in which the very survival of a State would be at stake".

Given the reckless and unlawful behaviour of the nuclear weapons states, an equivocation such as that in 2E must have left the nuclear weapons state laughing all the way to their nuclear armories! Such an ultra-statist view flies in the face of international humanitarian law and multiple treaty obligations. The judges cited no evidence for their view and clearly ignored the consequences of nuclear weapon use. No wonder Weeramantry produced a dissenting opinion more extensive that the ICJ Opinion itself. The ICJ Opinion can be found in full in NWIL3, Chapter 11, and Weeramantry's Dissenting Opinion in NWIL3 Chapter 12.

In the case of the Republic of the Marshall Islands (RMI) vs the world's nuclear weapons states, only three of those states were subject to the jurisdiction of the ICJ. Astonishingly, the ICJ dismissed the case on the grounds of the absence of a dispute. Further information on the RMI case is available in NWIL3 Chapter 13. see also Burroughs (1997).

Thus, in both of these substantial cases the ICJ has failed to carry out complete and thorough investigations There were extensive hearings and a mountain of documentation. The problem is not the lack of work, but their interpretation in the final judgment (para E) . It is highly regrettable that there is no mechanism to call for a formal review of ICJ decisions.

2.14 Renewed Mobilization of Civil Society

The failure by the NWS to comply with their obligations under the NPT, their failure to abide by many UNGA resolutions and the failures of the ICJ to carry out their deliberations completely and comprehensively, plus the stark reality set out in the ICJ Advisory Opinion at 2B (above) emphasizes the need for a "comprehensive and universal prohibition of the threat or use of nuclear weapons as such".

Civil society groups around the world took a few years to regroup for the challenges of establishing a comprehensive and universal prohibition of the threat or use of nuclear weapons - to which must be added 'possession', so the new goal is to establish law that prohibits comprehensively all aspects of nuclear weapons leading to the development, possession, threat to use, or use of nuclear weapons. Many groups remained active during this period. 1996-2007 was the time in which Abolition 2000 served as the main coordination for the abolitionist NGOs, and had many projects, notably the drafting of the NWC (see Chapter 4) - Abolition 2000 is still very active.

In 2007, several come together and formed ICAN - International Campaign to Abolish Nuclear Weapons.

ICAN and many other civil society groups and individuals continued to press for steps such as the TPNW.

2.15 Birth of TPNW

Starting in 2012 UNGA passed a series of resolutions for taking forward multilateral nuclear disarmament negotiations which finally resulted in a new treaty to ban nuclear weapons. They convey a sense of substantial frustration and determination by many countries because of that NPT failure by the nuclear weapons states.

The principal decision of UNGA Resolution 67/56 in 2012 was

> "... to establish an open-ended working group to develop proposals to take forward multilateral nuclear disarmament negotiations for the achievement and maintenance of a world without nuclear weapons".

The Open-Ended Working Group met in Geneva and submitted its report which was, by UNGA Resolution 68/46 in 2013, sent for further UNGA consideration.

Further consideration by UNGA in 2014 resulted in UNGA Resolution 69/41 which simply passed the matter on to the next UNGA session.

In 2015 UNGA considered the matter again and passed resolution 70/33. The principal decision of that resolution was to re-convene the OEWG in Geneva in 2016.

The OEWG met in 2016 and prepared its report in sufficient time to get it to the next meeting of UNGA later in 2016. That meeting later in 2016 resulted in UNGA Resolution 71/258 containing the critical decision:

> "*Decides* to convene in 2017 a United Nations conference to negotiate a legally binding instrument to prohibit nuclear weapons, leading towards their total elimination".

Participation was to be "with the participation and contribution of international organizations and civil society representatives".

More detail and text of the relevant UNGA resolutions can be found in NWIL3, Chapter 19.

After meeting at the UNGA in New York from 27 to 31 March and from 15 June to 7 July 2017, the negotiated TPNW was passed by a majority vote and became available for signing.

TPNW stipulated that it would enter into force 90 days after the 50th ratification of the Treaty. That happened on 22nd January,2021.

The nuclear weapons states played no part whatsoever in the TPNW negotiations.

2.16 Moving Forward from TPNW

Many people see TPNW as the beginning of the end of nuclear weapons. Let's hope they are right.

There is still a long way to go from TPNW to full multilateral nuclear disarmament.

After the TPNW entered into force, many civil society groups and TPNW-supporting states put forward a range of ideas to move the campaign forward. Here are some of them:

- Encourage more countries to sign, ratify, accede to, TPNW
- Ratifying and Respecting other Treaties and Agreements
- Alternatives to Deterrence and boosting the Green Economy (boosting the Green Economy is an effctive answerto objections based on unemployment)
- Nuclear Weapons Convention
- World Order, Nation States, Sovereignty and Peace - bring an end to the ability of sovereign states to hide behind their sovereignty to place the planet and humanity in peril;
- Create a Nuclear Weapons Register of Individuals and Organizations involved with Nuclear Weapons
- Persuade local authorities to support the Treaty and get prominent individuals to sign pledges of support.
- Persuade companies and banks to divest from nuclear projects on the basis that the TPNW renders them illegal.
- and...Be Vigilant and Do Something — each person can pick something to do to take forward TPNW;

2.17 Summary Timeline of the TPNW Journey

1930s - rapid increase in scientific research and development relevant to nuclear weapons.
1939-1945 - several countries embark on secret nuclear military programmes
1945 June - UN Charter signed
1945 July - 1st US nuclear bomb test
1945 August - 6th, Hiroshima atom bomb; 8th, Nuremberg Charter signed; 9th, Nagasaki atom bomb and the Soviet Union declares war on Japan
1945 October - UN Charter enters into force
1946 January - UNGA Resolution 1(1)
1948 - Israel starts its atom bomb research and development
1948 August - 1st Soviet Union atomic test
1952 - first British atomic test
1960 February - first French atom test
1960 and/or 1963 - first Israeli nuclear test
1964 October - first Chinese atom test
1968 - Non-Proliferation Treaty
1974 May - first Indian nuclear test
1998 May - first Pakistani nuclear test
2006 October - first North Korean nuclear test
2012-2016 - Open-Ended Working Group negotiations in Geneva pre-TPNW
2017 - TPMW (entered into force January 2021)

Nuclear Weapons Free Zones

1959 - Antarctic Treaty
1967 - Outer Space Treaty
1967 - Tlatelolco Treaty (Latin America)
1971 - Seabed Treaty
1979 - Moon Treaty
1985 - Rarotonga Treaty (South Pacific)
1987 - New Zealand Nuclear Free Zone
1995 - Bangkok Treaty (Southeast Asia); Pelindaba Treaty (Africa);
2000 - Mongolia (UN Security Council recognition as NWFZ)
2006 - Semipalatinsk Treaty (Central Asia)
2017 onwards - States Parties to TPNW

Chapter 3

Nuclear Weapons were illegal before TPNW

3.1 International Law

There is much public scepticism about the value of International Law. Many citizens and public officials continue to query whether international law is *really* law in the absence of police mechanisms for enforcement and procedures for impartial interpretation. The response here is that international law provides the underpinning for many varieties of transnational life that work so well we take it for granted. An effective legal order does not necessarily depend on central institutions for decision and enforcement. Law can be effective if the parties seek to make it so out of reasons of convenience, mutual benefit, a sense of right and respect, or even because they find value in a reputation of law–abidingness. These factors all operate to some extent in international life, varying from one substantive area to another, and from one kind of leadership to another.

Of course, all law is violated at times. Indeed, enforcement would be superfluous were compliance perfect. The special problem of international life and some domestic abuses such as genocide or actions against local populations, arises because some violations are so totally disruptive and unacceptable in their effects.

There is no doubt that considerations of reciprocity which ensure a high degree of effectiveness for international law (for example, upholding the immunity of foreign diplomats to reinforce the immunity of our own) are least operative in the context of war and peace where fundamental security, even survival, is at stake. Again, the wider context is important to appreciate. All law tends to bend and break in conditions of crisis, as is evident during periods of civil strife or economic privation.

Nevertheless, even governments have acknowledged over the centuries the great importance of bringing law to bear on decisions associated with *recourse to* and the *conduct of* war (*ad bellum* and *in bello*). From early times generally, and Grotius (see Grotius, 1682) in particular, onward there has been a consensus to the effect that unrestrained warfare was a regression to barbarism, unacceptable as such. Especially in the last century or so there has been a dual series of developments: first of all, the technological innovations in warfare, culminating in the development and use of atomic bombs and nuclear weapons; secondly, an intensifying insistence on restricting the discretion of governments to wage war "legally". Concerns have been extended in recent years to include non-kinetic forms of warfare, such as information warfare and cyber war (which must be assessed for lawfulness in accordance with Article 36 of Protocol 1 1977 to the Geneva Conventions, 1949 - no such assessments have been found by the Author although at a conference UK and USA government attendees claimed they had done such assessments - they are classified as 'secret'!).

There is no doubt that the possession of nuclear weapons underscores the tension between those political developments that give the modern state unconditional power over human destiny and those normative reactions by civil society (and many governments) that seek to impose limits upon what governments can do, even beneath the banners of military necessity and national security. Even the term 'national security' is often abused by using it to mean 'government security'. There is no doubt also, that since 1945 the statists have prevailed in relation to warfare and weaponry of mass destruction. There have been many wars and relatively little success in resolving conflict by recourse to the procedures made available by international law and embodied in the United Nations Organization. Of course,, the UN has helped prevent, resolve/end many conflicts). Today the statists prevail to such an extent that the current world order based on a collection of independent sovereign states has permitted several states to act as rogue states holding the planet and humanity to ransom by their nuclear weapons, even though most states are against them.

The leading nuclear weapons states have claimed that their possession of such weaponry has probably prevented World War III, and that the only reliable method to sustain "peace" is to threaten the annihilation of a rival society in retaliation. This system of mutual threat is generally called deterrence, and its logic and probable effects are not reconcilable with most understandings of law and morality. The nuclear tribunals (London and Australia) proceeded on the assumption that such a departure from normative restraint is dangerous and unacceptable, and that it does entail a relapse into barbarism on the grandest imaginable scale.

The nuclear tribunals (London and Australia) took cognizance of both sides of this modern dilemma. Firstly, the urgent need to replace deterrence with a system of international security responsive to law and morality. Secondly, the realization that governments and their institutional creations, including the United Nations, are not sufficiently motivated or empowered to satisfy this most fundamental of international needs. In these regards, the tribunals filled a normative vacuum. They intended to mobilize public opinion throughout the world around the necessity to bring available law to bear on the nuclear weapons policies of certain governments.

It is, at the same time, important to realize that the tribunals have not invented the legal framework they relied upon. This framework has been evolved over the years by governmental action responding to felt necessities and to the aspirations of the peoples of the world. We shall endeavour to demonstrate clearly that the international law interpreted and applied by the tribunals is of a status that should be applied by governments themselves.

In this respect, the tribunals were convened to fill a constitutional gap in the international political system of the present time. Their existence is an enactment of the call for individual responsibility (ie Nuremberg Principles) that is itself a signal achievement of modern international law and has been heralded as such by the main nuclear weapons states.

It is a startling irony that, aside from China, the other states that now acknowledge possession of nuclear weapons constituted the four prosecuting states at Nuremberg after World War II. In particular the two superpowers, the United States and the then Soviet Union, were most insistent that German leaders at all levels of society be held criminally liable for their refusal to uphold international law in the context of war and peace. The principles of the Nuremberg Charter continue today with the International Criminal Court (ICC).

The victorious governments were emphatic that their proceedings against the defeated governments of Germany and Japan would provide a framework for all political activities in international life. After the judgments had been given, a consensus among the victorious governments supported the effort to formulate the Nuremberg Principles as universally binding rules of international law. These Principles impose on governments and officials an unconditional duty to uphold international law regardless of state policy. The tribunals recognized that this duty serves the interests of all peoples, and that even the interest of one's own country is best upheld by assuring that its policies abroad conform to the rules of international law. It is then a matter of patriotic duty to insist on the application of the Nuremberg Principles, and most especially, in relation to the nuclear weaponry where so much is at stake and where it is too late to await an entire breakdown of order to establish the full evidence of a violations.

It is true that, at the time of the Nuremberg Judgment, and ever since, critics have dismissed the whole enterprise as "victors' justice". There is a sombre truth to this contention. As the Indian member of the Tokyo Tribunal, Justice Pal, pointed out, it was unacceptably hypocritical to accuse Japanese war leaders of crimes but exempt from scrutiny the Western indiscriminate bombing of Japanese cities climaxing in the atomic attacks on Hiroshima and Nagasaki. The tribunals acknowledged the imperfections in the legal precedents, but sought to build upon them to complete their promise.

It can also be alleged that the nuclear weapons states have not clearly accepted the view that these weapons are illegal. The tribunals considered carefully this allegation, but felt convinced by the evidence that international law exists with sufficient clarity to assess the policies of governments with respect to nuclear weapons. The tribunals agreed that a comprehensive treaty of prohibition would be a contribution to the avoidance of nuclear war, but that even without such a document, existing treaties and customary rules of international law are clear on these matters. Since the tribunals such a treaty has has now entered into force.(see also NWIL3 Chapter 19).

3.2 Substantive Foundations

The major sources of relevant international humanitarian law are treaties, customary law, and the public conscience (see NWIL3, 2020 for a detailed discussion of public conscience and the role of civil society in the evolution of applicable international law).

3.2.1 Just War Doctrine

Originating in theological discourse during the Middle Ages, the so-called Just War Doctrine was incorporated into positive international law. Drawing on still more ancient practices of political communities, the Just War Doctrine called upon participants in war to carry out combat with due respect for moral principles, including an overriding obligation to confine military attack to military targets, thereby avoiding any direct injury to civilians.

The application of the principles of the Just War Doctrine were left to each sovereign ruler, and amounted to an appeal to conscience. There was no higher authority aside from the questionable claim by the Roman Catholic Church that lost whatever overall validity it might have once possessed as a result of the fragmenting of Christendom due to the Reformation. In a sense, the Just War Doctrine was an instance of religious morality being converted into an applied ethics for international relations. It was borrowed and incorporated into international law at the inception of the state system.

The very creation of the Just War Doctrine happened because of tension between prior religious abhorrence of war and violence and the proclivities of those leaders who wished to enhance their power and assert their beliefs. Other religions have similar tensions. Although there are many religious sources of morality, there are inherent problems leading to barbarism in cases such as the status of non-believers. To see more about religious wars, religious aspects of colonialism and so forth see sources such as Richardson (1960) and Wright (1965).

3.2.2 Customary Norms of International Law

The lack of specificity in the Just War Doctrine was a shortcoming, as was the absence of any ritual of assent by which a sovereign authority acknowledged its duty to be bound in definite ways. International jurists collected the body of practices that governments accepted as binding upon themselves, and set forth these rules and principles under the rubric of customary international law. These rules and principles helped shape the direction of treaty law, and, as well, provided legally accepted yardsticks for measuring claims about the status of new weapons and tactics. Such rules and principles provide a normative background against which to evaluate the controversy about the lawfulness of nuclear weapons and about various doctrines governing their use.

o **Principle of Discrimination** - to be lawful, weapons and tactics must discriminate clearly between military and non-military targets, and be confined in their application to military targets. Indiscriminate warfare is per se illegal, although indirect damage to civilians and civilian targets is not necessarily so.
o **Principle of Proportionality** - to be lawful, weapons and tactics must be proportional to their military objective. Disproportionate weaponry and tactics are excessive, and as such, illegal.

- o **Principal of Lawfulness** - to be lawful, weapons and tactics must not violate any treaty rule of international law binding as between the parties.
- o **Principle of Necessity** - to be lawful, weapons and tactics involving the use of force must be reasonably necessary to the attainment of their military objective. No superfluous or excessive application of force is lawful, even if the damage done is confined to the environment.
- o **Principle of Humanity** - to be lawful, no weapon or tactic can be relied upon that causes unnecessary suffering to its victims, whether by way of prolonged or painful death, or in a form that is calculated to cause severe terror or fright. For this reason, weapons and tactics that spread poison, disease, or do genetic damage are generally illegal *per se*, as being weapons with effects not confined in the place and time of damage to the battlefield. Such a prohibition, under contemporary circumstances, extends to ecological disruption in any form.
- o **Principle of Neutrality** - to be lawful, no weapon or tactic can be relied upon that seems likely to do harm to human beings, property, or the natural environment in neutral countries. A country is neutral if its government declares itself to be so and if it pursues a policy of impartiality in relation to armed conflict, including the avoidance of any kind of alliance relationship.

To this list can be added new principles that have emerged in recent years, including principles incorporated into the Rome Statute:

- o **Principle of No Ecocide** - to be lawful, no weapon or tactic can be relied upon that seems likely to do serious harm to the natural environment.
- o **Principle of No Genocide** - to be lawful, no weapon or tactic can be relied upon that leads to the deliberate killing of a large number of people from a particular nation or ethnic group with the aim of destroying that nation or group.
- o **Principle of No Omnicide** - to be lawful, no weapon or tactic can be relied upon that is likely to do destroy all life or all human life.

3.2.3 Treaty Rules and Principles

Often it is assumed, wrongly, that international agreements in treaty form are the only valid source of international legal obligations. In some respects, written agreements, duly ratified, are preferable sources of guidance as to the requirements of international law. Written formulations can be more explicit and elaborate with respect to a given pattern of conduct. Furthermore, as far as governments are concerned, there is a tendency to accord greater respect to those legal obligations to which consent in explicit and constitutional form has been given, especially if the negotiation and ratification processes are recent, most particularly within the life span of the governmental leadership currently in power.

There are also limitations to the view that treaty rules are the only genuine source of international law, or even the more moderate view, that these formulations of law are necessarily the best source. Some general norms have not been reduced to treaty form. In other instances, some states are not bound by treaties, having withheld their consent. In still other instances, the content of the treaty rules is vague or subject to contradictory formulations, especially so in relation to the early efforts to codify war and peace, quite dramatically superseded by modern methods and styles of warfare, as well as by new military technologies and weapons systems.

There is an obvious problem of application with respect to nuclear weapons. The nuclear weapons states have so far refrained from entering into any serious negotiations towards a treaty, or even a declaration, acknowledging the unlawfulness of threats or uses of nuclear weapons. Such a deficiency is obviously not an oversight. Hence, to derive applicable rules of international law that add up to an unconditional prohibition of the use of this weaponry is bound to collide with the official security policies of major states, and to challenge the legitimacy of weapons capabilities and bureaucracies that command control over vast allocations of resources.

See the sister book to this one for a more detailed discussion of relevant treaties (Darnton et al., 2020). The important general treaties in this area were formulated at the Hague in 1899 and 1907 in a series of comprehensive conventions that summarized the pre-World War I levels of agreement as they existed between the governments playing a leading role in international life. The goal was not to eliminate war, but to regulate its conduct in accordance with the customary principles briefly set forth in the preceding paragraph. Especially important was the broad imperative embodied as a common article in the various Hague Conventions of 1899. Article 22 in the Annex to the Hague Convention IV (Regulations Respecting the Laws and Customs of War on Land) states:

> "The right of belligerents to adopt means of injuring the enemy is not unlimited"

The apocalyptic implications of a major reliance on nuclear weapons gives this provision an obvious orienting relevance.

Also critical was the celebrated "de Martens clause" (named after the Belgian jurist Feodor de Martens) inserted in the 1907 Hague Conventions[1]:

> "Until a more complete code of the laws of war has been issued, the high contracting Parties deem it expedient to declare that, in cases not included in the Regulations adopted by them, the inhabitants and belligerents remain under the protection and the rule of the principles of the law of nations, as they result from the usages established among civilised people, from the laws of humanity, and the dictates of the public conscience".

This resolve in international treaty law to base permissible action expressly on normative traditions and upon conscience is significant. The "de Martens clause" definitely refutes the ultra-statist view that everything is permitted if it has not been expressly renounced by a formal manifestation of governmental authority.

International treaty law has successfully achieved a very widely endorsed prohibition of poison as a weapon and tactic of war. To date the most important treaty instrument, adopted in response to the menace of poison gas revealed in the trenches of World War I, is the 1925 Geneva Protocol for the Prohibition of the Use in War of Asphyxiating, Poisonous and other Gases, and of Biological methods of Warfare.

Following on from the 1925 Geneva Gas Protocol, a Biological Weapons Convention (BWC, 1972) was negotiated and made available for signature on 10th April 1972 and entered into effect on 26th April 1975. A Chemical Weapons Convention (CWC, 1993) opened for signature on 19th January 1993 and entered into force on 29th August 1997. The absence of a Nuclear Weapons Convention has long been relied on by NWS as an effort to legitimize their nuclear weapons, hence this book includes material for a Nuclear Weapons Convention (NWC, 2007; Appendix B)

Another significant line of effort in treaty law concerns the discretion to initiate war via acts of aggression. In the 1928 Pact of Paris, war is outlawed as an instrument of national policy, and legitimate force confined to circumstances of self-defence. This treaty norm provided a major basis for the war crimes prosecution at Nuremberg and Tokyo after World War II, giving rise to the category of offence known as "Crimes against Peace". The United Nations Charter, a multilateral treaty, carries forward in Articles 1(4), 33, and 51 the basic notion that there is no legal pretext for recourse to force in international

1 From the preamble to the 1907 Hague Convention IV Respecting the Laws and Customs of War on Land.

relations except in self-defence against a prior armed attack. There is some controversy among international law specialists as to whether patterns of state practice have so consistently ignored this constraining legal framework as to suspend, or to draw into question, its continuing validity. At stake in the nuclear weapons setting is the critical issue as to whether the design and development of first-strike weaponry and supporting doctrine amounts to a *per se* act of aggression, as well as rendering officials liable for crimes against the peace. At Nuremberg it was definitely decided that planning for aggressive war is itself a crime even if the aggressive policy is never consummated. Does this prohibition pertain to those allegations that certain classes of nuclear weapons systems have first-strike properties and roles? Clearly it does.

Another major treaty instrument was the Genocide Convention of 1948 that established the criminality of any course of deliberate state policy that intends to destroy, in whole or part, national, ethnic, religious, or racial groups. Nuclear weapons are aptly described as weapons of mass destruction, and their use would be genocidal in impact, as well as ecocidal. Indeed, the grim magnitude of such destruction suggests that beyond genocide lies the result of omnicide.

Ever since the nineteenth century there has been an effort complementary to that of the Law of the Hague dealing with weapons and tactics to codify international humanitarian law applicable during wartime, sometimes known as the Law of Geneva, because so many of the main treaty instruments were negotiated and signed at Geneva. The main elements of the Law of Geneva are the four Geneva Conventions of 1949 — for the protection of land forces, of sea forces, of prisoners of war, and of civilians. The attempt of these agreements is to give concrete application to the Principle of Humanity, by imposing obligations on belligerent states to respect the sanctity of such things as hospitals, cultural monuments etc., and to avoid any military action against the sick and wounded, or against those of the enemy who have laid down their arms and become prisoners of war.

These treaty rules suggest levels of respect for the limits of warfare that seem utterly inconsistent with any use of nuclear weapons.

Both the United States and Britain made it clear that their participation in the negotiation of the Geneva Protocols I and II in 1977, to modernise the 1949 Conventions, was taking place on the assumption that nuclear weapons were not to be considered subject to the treaty norms, even in relation to Article 35 which explicitly deals with new weapons and methods of warfare. Is such an exclusion effective?

A final source of treaty guidance arises from the legal duty imposed on the governments of nuclear states by such arms control agreements as

32

the Limited Test Ban Treaty of 1963 and the Non-Proliferation Treaty of 1968, to negotiate in good faith an end to the nuclear arms race and to establish by stages or any reasonable process, secure arrangements for general and complete disarmament.

3.2.4 Supplementary Norms with Legal Force

The United Nations has itself contributed in a variety of respects to the development of international law. General Assembly resolutions have been claimed to have a limited legislative effect under certain conditions of their passage. The UN General Assembly has manifested its concern about the lawful status of nuclear weapons in a long series of widely endorsed resolutions going back to General Assembly Resolution 1653(XVI) which clearly supported the view that threats or uses of nuclear weapons were violations of the UN Charter and constituted Crimes against Humanity. The United States and its NATO allies voted against this and other subsequent resolutions on this subject matter.

A closely related concern involves the status of initiating use of nuclear weapons. Both the Soviet Union and China made in 1981, a unilateral and unconditional commitment never to use nuclear weapons first. The Western nuclear powers have not acceded in any formal way to this no-first-use position. What status the no first use position has in contemporary international law is an important issue to be examined. No first use measures and agreements are useful Step by Step measures as were the two test ban treaties

Natural law criteria of state behaviour are also applicable through a continued reliance on the de Martens clause, and its invocation of the "laws of humanity" and "the dictates of the public conscience". Such a moral outreach *within international law* makes it important and entirely appropriate to consider for *legal relevance* the great variety of statements by religious bodies describing their urgent concern and supporting reasoning about the irreconcilability of current doctrines pertaining to the use of nuclear weapons and the dictates of public conscience. Such an assessment is reinforced by a growing number of independent experts lending their professional judgment to the view that current policies of nuclear weapons states violate international law in flagrant and serious ways and to varying degrees. Such materials by religious bodies or international jurists are not law as such, but *evidence* as to the content of law, especially given the legal duty by governments to respect the dictates of public conscience in their war-making activities. A large manifestation of the public conscience was the millions of signatures delivered by the World Court Project to the ICJ when considering its 1996 Advisory opinion.

3.3 Nuremberg Principles of International Law

Principle I — Any person who commits an act which constitutes a crime under international law is responsible therefore and liable to punishment.

Principle II — The fact that internal law does not impose a penalty for an act which constitutes a crime under international law does not relieve the person who committed the act from responsibility under international law.

Principle III — The fact that a person who committed an act which constitutes a crime under international law acted as Head of State or responsible government official does not relieve him from responsibility under international law.

Principle IV — The fact that a person acted pursuant to order of his government or of a superior does not relieve him from responsibility under international law, provided a moral choice was in fact possible to him.

Principle V — Any person charged with a crime under international law has the right to a fair trial on the facts and law.

Principle VI — The crimes hereinafter set out are punishable as crimes under international law:

a. **Crimes against peace:**
   i Planning, preparation, initiation or waging of a war of aggression or a war in violation of international treaties, agreements or assurances;
   ii Participation in a common plan or conspiracy for the accomplishment of any of the acts mentioned under (i)

b. **War Crimes:**
   Violations of the laws or customs of war which include, but are not limited to, murder, ill-treatment of prisoners of war or persons on the seas, killing of hostages, plunder of public or private property, wanton destruction of cities, towns, or villages, or devastation not justified by military necessity.

c. **Crimes against humanity:**
   Murder, extermination, enslavement, deportation and other inhuman acts done against any civilian population, or persecutions on political, racial or religious grounds, when such acts are done or such persecutions are carried out in execution of or in connection with any crime against peace or any war crime.

Principle VII — Complicity in the commission of a crime against peace, a war crime, or a crime against humanity as set forth in Principle VI is a crime under international law.

3.4 Nuclear Deterrence is Unlawful

The use of nuclear weapons is unlawful Therefore nuclear deterrence, which is the threat to use an unlawful weapon, is unlawful also.

For a more detailed discussion of this, see Boyle (1986: 2002) ,Drummond (2019) and Kahan (1999) . Boyle was a participant in the LNWT so more information about him is available in NWIL3 and Wikipedia.

3.5 Conclusions

The net conclusion to be drawn from customary and treaty law summarized above, combined with the dictates of the public conscience, is that nuclear weapons have been unlawful since the very beginning of their existence, well before TPNW. Taking the Nuremberg Principles into account TPNW consolidates the law and names nuclear weapons specifically.

Cases where nuclear weapons states have stated reservations in the applicability of some treaty provisions (such as the UK and USA reservations with respect to the 1977 Geneva Convention Protocols, or various reservations about nuclear free zones, and the Rome Statute — ICC) need to be noted, but are ineffective due to the NPT, UN Charter, and customary international law. Some writers hold that the Non-Proliferation Treaty 1968 legitimates some nuclear weapons states - it does no such thing, as it merely acknowledges their existence. It does not legitimate their nuclear weapons.

Unlawfulness of nuclear weapons goes as far back as the bombing of Hiroshima and Nagasaki. For am early legal judgment finding the bombings to be unlawful, see the Shimoda case (Shimoda, 1963; Falk, 1965). The geopolitical reasons for those bombings go nowhere near the very remote threshold suggested as unknown in the International Court of Justice Advisory Opinion, 1996 (see Ch 2).

Not only have nuclear weapons been illegal since the birth of the bomb, but also the Nuremberg Principles mean that all individuals and legal entities that have been involved in decisions to develop and deploy nuclear weapons along with their manufacture, financing, ,maintenance and so forth, become individually and personally liable under international criminal law.

Throughout the whole evolution of customary and treaty international law, civil society has played pivotal roles in encouraging, developing, and drafting. Grotius, the founding father of international law, can be seen as a highly motivated civil society advocate. More recently, at the time of the Hague Conventions, and subsequently, the International Peace Bureau (IPB) and its partners have been key (see Archer 2020). Civil Society will remain at the heart of moving the international community forward from TPNW towards nuclear abolition.

What TPNW Does and Does Not Do; TPNW vs NPT vs NWC

4.1 Introduction

TPNW can be seen as the beginning of the end of nuclear weapons. A majority (122) of the world's countries voted for it it in UNGA; the nuclear weapons countries ignored the TPNW negotiations (notwithstanding NPT obligations) but they cannot ignore that it has entered into force. The noose is tightening around the development and possession of nuclear weapons.

The essence of this book is to explore ideas for taking TPNW forward. To start that process it is important to understand what TPNW does, and what it does not do. Therefore, read this Chapter in conjunction with Appendix A and other sources.

The principal weapons of mass destruction are generally considered to be biological, chemical, and, nuclear. Conventions already exist for biological (BWC) and chemical weapons (CWC). The fact that there is not yet a nuclear weapons convention (NWC) has been exploited many time by nuclear weapons states asserting that nuclear weapons are not explicitly prohibited *per se* in international law. That position is an extreme but invalid assertion by nuclear weapons states. A useful discussion of this point can be found in Lippman (1986) For further discussion of the Lotus case and an early discussion of the need for a nuclear weapons legal regime, see Falk (1983). Of course, whatever the nuclear weapons states say to justify their position that nuclear weapons are not prohibited *explicitly* as such by international treaty law, these states are subjects to customary international law and various treaties whatever types of weapons or war they talk about.

The absence of a nuclear weapons convention, and the over-reliance on that fact by nuclear weapons states, was an important factor encouraging the development of TPNW. A major point of TPNW is that it does not go as far as the proposed NWC, therefore pressure for a NWC continues. An early draft NWC can be found in Datan and Ware (1999), updated in Datan et al. (2007). Appendix B shows a draft

NWC as presented to UNGA (NWC, 2007). Read Appendix B also, in conjunction with this Chapter..

This chapter includes discussion of differences between TPNW and the proposed NWC in Appendix B.

4.2 What TPNW Does

TPNW is focussed primarily on nuclear weapons and nuclear explosive devices. That is clear from the list of prohibitions set out in Article 1. This means it is not concerned about what some people refer to as 'peaceful uses' of nuclear energy. Thus, some dual uses can escape the effects of the Treaty, such as nuclear propulsion, and many nuclear reactors. This matters for taking forward the development of the Treaty.

The prohibitions are wide-ranging, including: do not develop, test, produce, manufacture, otherwise acquire, possess or stockpile nuclear weapons or other nuclear explosive devices; do not transfer to or from anywhere nuclear weapons or other nuclear explosive devices or control over such weapons or explosive devices directly or indirectly; do not use or threaten to use nuclear weapons or other nuclear explosive devices; do not assist, encourage or induce, in any way, anyone to engage in any activity prohibited to a State Party under this Treaty; similarly, none of these should be received; do not allow any stationing, installation or deployment of any nuclear weapons or other nuclear explosive devices in your territory or at any place under your jurisdiction or control.

The Treaty provides for mechanisms to be used in the case of a party to the Treaty being involved with nuclear materials or activities to ensure cessation and non-diversion of nuclear materials from peaceful activities.

States must ensure the removal from their territory of any nuclear weapons or explosive devices of any other state.

Depending on the interpretation of 'facility' the Treaty could require States Parties to make any institution on its territory that is involved in activities such as financing or researching nuclear weapons or nuclear explosive devices to cease all such activities and obtain verification that that has happened.

States should enact local legislation and introduce penal sanctions against anyone who breaches the requirements of the Treaty.

States Parties must provide victim assistance and environmental remediation for any effects of testing nuclear weapons or nuclear explosive devices.

States Parties are to assist each other in providing victim support and environmental remediation, particularly states that have caused victims or environmental degradation in other states as a result of testing nuclear weapons or nuclear explosive devices.

Thus, TPNW provides a framework for the total elimination of nuclear weapons and explosive nuclear devices if sufficient countries join and implement it.

It is couched in rather narrow terms and is certainly capable of being wider and more comprehensive. It could be the beginning of the end of nuclear weapons if taken forward.

TPNW requires implementation in domestic law (Article 5).

4.3 What TPNW Does Not Do

TPNW is focussed primarily on nuclear weapons and nuclear explosive devices. It does not discuss nuclear facilities or nuclear weapon delivery vehicles.

Although TPNW does require national implementation by States Parties, the requirements are weak and vague. Far more detailed provisions would help to achieve comprehensive implementation.

Nuclear weapons and their supply chains from basic resources through to possession, deployment, and use are extremely expensive in terms of finance and resource usage. TPNW is silent on these which could only be implied from the requirement for national implementation. There need to be explicit prohibitions on activities such as: providing finance and investment for nuclear weapons; carrying out research or advanced development for nuclear weapons activities; education and training of people to be involved with nuclear weapons activities (apart from verifying, dismantling, re-deploying or disposing of nuclear weapons materials or facilities). Further information about this can be found in Chapter 18 of NWIL3 — Move the Nuclear Weapons Money — and the website with the same title[1] which is independent of this book and NWIL3 but provided inspiration for that Chapter title.

The Nuremberg Principles and Rome Statute (the basis of the International Criminal Court - ICC) exist independently and prior to the TPNW. These are concerned with international criminal law applying to individuals. TPNW is largely silent on these in relation to nuclear weapons, explosive nuclear devices and nuclear weapons facilities or activities. There is a need for a more detailed explanation

---

1 http://www.nuclearweaponsmoney.org/

of the application of the Nuremberg Principles and ICC to individuals involved with all activities related to nuclear weapons and nuclear explosive devices. This is discussed in more detail in NWIL3.

A major topic that is used so often as a cloak for secret nuclear weapons activity is what is called euphemistically 'peaceful uses of nuclear energy'. So many negotiations about nuclear weapons and nuclear explosive devices move quickly in inserting clauses permitting 'peaceful' uses (and the Model NWC is a prime example). The pressures to eliminate nuclear weapons are because such weapons pose an immediate existential threat to the world and humanity. This book is being written in the middle of the Covid-19 pandemic seeing all round the world the capability of the virus to place intolerable strains on medical services everywhere. Only one, or a very small number of exploded nuclear weapons would pose challenges to medical services that would make the challenges from Covid-19 pale into insignificance.

As soon as the time horizon for the effects of nuclear weapons activities changes from immediate to medium- or long-term, the whole of nuclear energy activity comes within the scope of existential threats to the world and humanity. The fundamental and key problem of rendering nuclear waste to be safe has not been solved. Stockpiles of nuclear waste are building up in many parts of the world. This situation (inability to render nuclear waste to be safe) poses a major problem for nuclear disarmament: nuclear weapons radioactive material would need to be added to existing stockpiles.

Some nuclear isotopes are so dangerous for so long, that storage will require many thousands of years (for example, the half-life of Plutonium is over 24 thousand years; Uranium 235,about 700 million years)) . This imposes a huge liability on many future generations. For the long-term storage of nuclear waste, the data simply does not exist to know the effects of climate change on storage facilities, or to know which geological formations are or will be stable enough over thousands of years or millennia. The nuclear waste from one nuclear power station will blight humanity for thousands of years - for a facility that may exist for only 40-60 years. When nuclear power station investment decisions are taken, the calculations do not take account of the thousands of years or millennia needed for waste storage when making cost-benefit calculations.

The radiation effects of the bombings of Hiroshima and Nagasaki took the US scientists by surprise; they had not been anticipated. Subsequent nuclear testing in regions such as the South Pacific[2],

---

2 For a musical film presenting information and emotion about South Pacific

Moruroa[3], Semipalatinsk and other test sites have resulted in a substantial increase in ubiquitous environmental radiation that is entering food chains. For an example of a multi agency analysis of radiation in food and the environment, conducted by a number of agencies of the UK's devolved governments, see RIFE (2019).

Now that radiation is ubiquitous around the globe, when scientists want materials with no radiation they need to use sources that existed before 1945, such as water from glaciers, or steel recovered from ships sunk before 1945. This all means that the narrative about 'peaceful' uses of nuclear energy needs to change a lot when negotiating the prohibition of nuclear weapons; many so-called peaceful uses of nuclear energy need to be brought within the scope of prohibitions. This also means that research into the safe disposal of nuclear waste is important to facilitate nuclear disarmament and the disposal of nuclear materials from nuclear weapons, as well as waste from nuclear energy uses more generally.

The establishment of 'research' nuclear reactors has always been a precursor to the development of nuclear weapons, takes a country much nearer to a nuclear weapons development capability. Any economic argument in favour of nuclear power stations has gone away as producing electricity from renewable resources is much cheaper than from nuclear power stations. A nuclear power station development programme is always masking the production of nuclear materials for some other purpose clouded by government deception and secrecy.

In order to have a clear view of the differences between what TPNW does and does not do, the final section of this Chapter presents a summary of the structures and contents of TPNW and the 2007 UNGA Model NWC.

4.4 Critics of TPNW

All the discussions and negotiations between 2012-2017 leading up to TPNW were plagued by the absence of all the nuclear weapons states. Some states that can be called nuclear umbrella states, linked to one of the nuclear weapons states (the largest number being NATO states) did participate in the OEWG discussions and negotiations but finally voted against. Interestingly, according to the UN voting record, China, India, and Pakistan abstained from voting on the UNGA resolution

---

tests, see the award winning From Scratch (1993).
3 For an impressive presentation about the effects of testing at Moruroa, see Moruroa (2021) — Moruroa-files.org.

71/258 setting up the negotiations for TPNW.

The voting record on the negotiated TPNW can be found at the end of Chapter 19 in NWIL3.

So what are the likely reasons for voting against TPNW or not taking part in the negotiations, particularly given that most countries are Sates Parties to the NPT? The likely reasons range from the ultra-statists who operate a realpolitik that as sovereign states, if they want nuclear weapons, nobody else can do anything about it, via manufactured mythology that TPNW would undermine the NPT, and/or deterrence theory, to the simple reality that some states decided to avoid the costs of participation knowing that they can vote without being a participant.

4.4.1 Realpolitik

The nuclear weapons states terrorize the world with their nuclear weapons in spite of their various international legal obligations and existing international law. They are therefore technically the largest terrorist organizations on the planet notwithstanding their various condemnations of terrorism.

What they are doing is spectacular hypocrisy of some of the NWS who also behave as imperialists and conducting or supporting foreign wars, exercising what they see as their sovereign rights under the current world order to do as they please notwithstanding the legal obligations they have taken on themselves willingly. None has demonstrated it can be trusted with respect to legal obligations willingly taken on and all show contempt for international law. They all act or have acted as bullies on the international stage. This is the *Realpolitik* of the NWS.

4.4.2 TPNW vs NPT?

During the discussions and negotiations leading to TPNW one theme recurred many times: TPNW could undermine the NPT. This was the approach taken by NATO (see Appendix D for a news brief giving such a summary). As a result NATO states and other states operating under a nuclear weapons umbrella voted against TPNW.

The fundamental problem with the information operations performed by NATO and their member states against TPNW is that they have never been accompanied by any comprehensible explanation as to *how* TPNW could undermine the NPT. Also, it is noteworthy that NATO members have not complained to the nuclear weapons states that are States Parties to the NPT asking why they remain in breach of the legal obligations under NPT Article VI. More to the point, the members of

NATO who are also States Party to the NPT also have the same legal obligation under NPT Article VI - why are they not banging on the doors of the nuclear weapons states to start multilateral negotiations in good faith for complete nuclear disarmament? It would be far more satisfactory if NATO and its members provided a rational explanation of how TPNW could undermine NPT rather than indulging in naive information warfare operations.

It is not only NATO countries that indulged in supporting assertions that TPNW would undermine the NPT; in an extremely rare act of cooperation, the UNSC P5 issued a document to the same affect (see P5). Key documents are in Appendices D-F.

International legal experts would welcome the opportunity to have some to scrutinize about the alleged ways TPNW could undermine NPT.

For the debate about TPNW vs NPT, it is worth noting the abstract of a paper prepared by a member of the Disarmament, Arms Control and Non-Proliferation, The Austrian Federal Ministry for Europe and International Affairs, Vienna, Austria:

> "Great care was taken during the negotiations of the Treaty on the Prohibition of Nuclear Weapons (TPNW) to secure its full compatibility with the Treaty on the Non-Proliferation of Nuclear Weapons (NPT). This goal has been accomplished. The TPNW strengthens and supports the NPT which has always anticipated further legal norms to achieve its purposes. Like in the other pillars of the NPT, reaching the objective of the disarmament pillar – a world free of nuclear weapons – will not be feasible without further legal instruments. For the full implementation of Article VI of the NPT, the creation of a legally binding norm to prohibit nuclear weapons is indispensable. The adoption of the TPNW on 7 July 2017 brought about this legal instrument." (Hajnoczi, 2020).

It is also worth noting that NPT is also concerned primarily with nuclear weapons and explosive nuclear devices. It can be argued that the NPT's glowing advocacy and encouragement of developing the 'peaceful' uses of nuclear energy can have an effect contrary to the key purposes of the NPT by helping more countries to move nearer to a nuclear weapons capability. Perhaps that was one price to pay for bringing some non-nuclear-weapons counties into the fold of the NPT.

Therefore, until such time as someone can come up with a convincing argument about how TPNW can undermine NPT, TPNW must be seen as complementary to NPT not undermining of it.

Those nuclear weapons states that have added reservations to their accessions to any NWFZs are probably in breach of Article VII of the NPT.

4.4.3 Deterrence Theory

Many writers about deterrence refer to deterrence theory. If deterrence in a context of nuclear weapons is the topic, what is the theory?

The scope of deterrence theory is far wider than nuclear weapons and has been explored in fields such as child development and behaviour, crime, and war generally. The core concept of deterrence is that certain behaviours will or may attract punishment and that the punishment acts as a deterrent of certain behaviours. There are related issue such as detection of the behaviours, certainty of punishment, balancing the costs of offending behaviour and costs of punishment.

For those interested in deterrence theory, have a look at Kuo et al. (2020) for a meta-analysis of deterrence theory in a different context. The hypotheses established in that meta-analysis such as "H1.1 There is a significant positive relationship between punishment severity and security-compliant behavior" (ibid p4) could be applied in an international relations context.

There has been no similar deterrence theory analysis carried out in the international relations field generally or nuclear weapons in particular. Interestingly from the Kuo et al. meta-analysis, the results for H1.1 shown above was only weak. The strongest relationship was related to subjective norms, with the next strongest relationships (moderate) for factors such as moral beliefs. Thus, those individuals who are given the responsibility with nuclear weapons to 'push the button' may well decline to do so, such as the Soviet Union submarine senior officer (Vasili Arkhipov,) during the Cuba crisis (Davis, 2017; Savranskaya, 2005)..

This suggests that a medium- to long-term goal of civil society should focus on making individuals aware of the moral problems of nuclear weapons and individual liability for individuals according to the Nuremberg Principles and Rome Charter.

A major problem with deterrence theory in the context of nuclear weapons is that deterrence is concerned with punishment and retaliation. There is no place in international law for punishment or retaliation. Self-defence may be ok but is subject to international law that would exclude the use of nuclear weapons.

Therefore, nuclear deterrence is illegal under international law - not that the leaders of the nuclear weapons states are likely to care about that - there are far too many leaders who enjoy possessing their big toys! All those leaders and military personnel involved with nuclear weapons need to be identified and outed. This is not simply a state problem; it is a legal problem for all involved individuals.

An example can be explored looking at the UK which claims to have an independent nuclear deterrent in the form of submarine launched Trident missiles. The missiles are made in the USA so they are hardly UK independent missiles.

Those interested in reasons against Trident should consult sources such as Green (2018) and Forsyth (2020). A definitive discussion about the illegality of deterrence is Boyle (2002).

4.4.4 Costs Burden

Some states did not participate in the OEWG or final TPNW negotiations simply not to have the costs of participation, knowing that they would be able to vote for, against, or abstain in the final UNGA vote.

4.5 TPNW vs Model NWC

This section presents summaries of the structure and contents of TPNW and the 2007 proposed NWC (full texts in Appendices A & B).

It is easy to see that not only is the NWC substantially more comprehensive, but also where there seem to be corresponding Articles, the NWC Articles are usually much more detailed than those in the TPNW.

Definitions are far more comprehensive and detailed in NWC than in TPNW.

Now it is 2021 and the Model NWC was produced in 2007, a first step in negotiating a NWC would be to correct it and bring it up to date.

4.5.1 TPNW Contents

Article 1 - Prohibitions; Article 2 - Declarations;Article 3 - Safeguards Article 4 - Towards the total elimination of nuclear weapon; Article 5 - National implementation; Article 6 - Victim assistance and environmental remediation; Article 7 - International cooperation and assistance ; Article 8 - Meeting of States Parties; Article 9 - Costs; Article 10 - Amendments; Article 1 - Settlement of disputes; Article 12 - Universality; Article 13 - Signature; Article 14 - Ratification, acceptance, approval or accession ; Article 15 - Entry into force; Article 16 - Reservations; Article 17 - Duration and withdrawal; Article 18 - Relationship with other agreements; Article 19 - Depositary; Article 20 Authentic texts

44

4.5.2 Model NWC Contents

<h2>Summary of the Model Nuclear Weapons Convention</h2>

I. General Obligations II. Declarations III. Declarations
IV. Phases for Implementation V. Verification
VI. National Implementation Measures
VII. Rights and Obligations of Persons VIII. Agency
IX. Nuclear Weapons X. Nuclear Material XI. Nuclear Facilities
XII. Nuclear Weapons Delivery Vehicles
XIII. Activities Not Prohibited Under This Convention
XIV. Cooperation, Compliance and Dispute Settlement
XV. Entry Into Force XVI. Financing XVII. Amendments
XVIII. Scope and Application of Convention
XIX. Conclusion of Convention
Optional Protocol Concerning the Compulsory Settlement of Disputes
Optional Protocol Concerning Energy Assistance
Annex I. Nuclear Activities; Annex II. Nuclear Weapon Component;s
Annex III. List of countries and geographical regions for the purpose
of Article VIII.C.23; Annex IV. List of countries with nuclear power
reactors; Annex V. List of countries and geographical regions with
nuclear power reactors and/or nuclear research reactors

4.6 Nucler Terrorism Convention

The Nuclear Terrorism Convention (NTC, 2005), signed by all the
NWS except North Korea and Pakistan would have identified all
relevant individuals in the NWS as responsible for terrorism had the
NWS not insisted in the insertion of Article 4(4):

> "4.This Convention does not address, nor can it be interpreted as addressing,
> in any way, the issue of the legality of the use or threat of use of nuclear
> weapons by States"

However this may be inapplicable; the convention is couched in
terms of individuals. It is too anthropomorphic to talk about Staes
using or thretening to use nuclear weapons; it is individuals who do
that. Individuals must refuse to use nuclear weapons. For examples
of imdividual military officers unwilling to use nuclear weapons, see
Davis,2017; Green, 2018; Forsyth, 2020; and, Savranskaya, 2005.

Lawyers would have a filed day arguing this one, but a fundamental
rule of the construction of statutes and deeds that the plain use of the
words must be used, and once something is enacted, it is too Late to
argue that the drafters ment something else (Bennion, 2008).

Chapter 5

The Importance of Domestic Prohibition Measures: the NZ Example

5.1 Introduction

Customary international law constrains sovereignty. States willingly accept further constraints when they enter into international treaties (for example, the UN Charter). Some states accept constraints, then cynically fail to act on them (the examples of key relevance to this book are the nuclear weapons states who are parties to the NPT but for over 50 years have failed to meet their obligations under NPT Article VI).

In terms of the relationship between international and domestic law, in some cases states *may* implement treaty obligations in domestic law; in some cases a state *must* implement treaty obligations in domestic law; and in some cases *it would be better* if countries implemented international legal obligations in domestic law, but usually do not (an example of this of direct relevance to this book is the Nuremberg Principles).

This chapter takes a brief look at the requirements for implementation in domestic law for states parties to the Treaty on the Prohibition of Nuclear Weapons (TPNW). The TPNW does not include any domestic legislation template for implementation in domestic law. One English-speaking state (examples in other languages can be submitted for inclusion in this book's website) that has implemented domestic legislation, although not required to do so by any treaty, is New Zealand. Also New Zealanders were very active in diplomatic work to obtain the 1996 Advisory Opinion (AO) from the International Court of Justice (ICJ) about nuclear weapons, and the development of the TPNW itself. Therefore the book also takes a brief look at that NZ domestic legislation.

New Zealand is a State Party to the TPNW so is doubtless in the process of drafting its required domestic legislation. It may well be able to do so by amendment of its 1987 Nuclear Free Zone, Disarmament and Arms Control Act to cover the requirements of incorporating the TPNW into domestic legislation.

5.2 TPNW Requirements for Domestic Legislation

TPNW Article 5 (see Appendix A) provides that:

Article 5 - National implementation

1. Each State Party shall adopt the necessary measures to implement its obligations under this Treaty.

2. Each State Party shall take all appropriate legal, administrative and other measures, including the imposition of penal sanctions, to prevent and suppress any activity prohibited to a State Party under this Treaty undertaken by persons or on territory under its jurisdiction or control.

The necessary domestic legislation must address at a minimum, the prohibitions listed in the Treaty:

Article 1 - Prohibitions

1. Each State Party undertakes never under any circumstances to:

(a) Develop, test, produce, manufacture, otherwise acquire, possess or stockpile nuclear weapons or other nuclear explosive devices;

(b) Transfer to any recipient whatsoever nuclear weapons or other nuclear explosive devices or control over such weapons or explosive devices directly or indirectly;

(c) Receive the transfer of or control over nuclear weapons or other nuclear explosive devices directly or indirectly;

(d) Use or threaten to use nuclear weapons or other nuclear explosive devices;

(e) Assist, encourage or induce, in any way, anyone to engage in any activity prohibited to a State Party under this Treaty;

(f) Seek or receive any assistance, in any way, from anyone to engage in any activity prohibited to a State Party under this Treaty;

(g) Allow any stationing, installation or deployment of any nuclear weapons or other nuclear explosive devices in its territory or at any place under its jurisdiction or control.

The prohibitions over possessing or transferring nuclear weapons or explosive devices is largely a state matter, but domestic legislation should prevent a state from reserving any prerogative powers that can be exercised 'secretly'.

The key prohibitions to be applied to individuals and other legal entities are those set out in clause 1(a) and (e) above.

The domestic legislation can prohibit individuals and legal entities from all activities involved in developing, deploying, using, maintaining, or financing related to nuclear explosive devices, anywhere in the world.

A weakness in the TPNW is that it covers only nuclear weapons and other nuclear explosive devices - it does not cover other nuclear materials or nuclear weapon delivery systems. However, domestic legislation could plug those holes.

5.3 New Zealand Domestic Legislation

New Zealand is a State Party to the South Pacific Nuclear Free Zone (SPNFZ) Treaty of Rarotonga, 1985 (see Rarotonga, 1985). This Treaty has no obligations similar to the TPNW to enact domestic legislation as part of implementing the latter. Therefore, the decision by New Zealand to implement domestic legislation was a unilateral act following its ratification of the SPNFZ Treaty. New Zealand's strong anti- nuclear weapons policy followed a period of civil society activities, reactions against New Zealand's involvement in the Vietnam War, French nuclear tests in the South Pacific, and with a receptive government in power from 1984-1990.

As already mentioned in section 5.1 above, New Zealand's domestic legislation is the New Zealand Nuclear Free Zone, Disarmament, and Arms Control Act 1987 (see Appendix C). Having just implemented the SPNFZ Treaty, this ground-breaking national legislation did adopt its preambular goals.

The reprint notes to the Act explain:

"An Act to establish in New Zealand a Nuclear Free Zone, to promote and encourage an active and effective contribution by New Zealand to the essential process of disarmament and international arms control, and to implement in New Zealand the following treaties:
(a) the South Pacific Nuclear Free Zone Treaty of 6 August 1985 (the text of which is set out in Schedule 1):
(b) the Treaty Banning Nuclear Weapon Tests in the Atmosphere, in Outer Space and Under Water of 5 August 1963 (the text of which is set out in Schedule 2):
(c) the Treaty on the Non-Proliferation of Nuclear Weapons of 1 July 1968 (the text of which is set out in Schedule 3):
(d) the Treaty on the Prohibition of the Emplacement of Nuclear Weapons and Other Weapons of Mass Destruction on the Sea-bed and the Ocean floor and in the Subsoil Thereof of 11 February 1971 (the text of which is set out in Schedule 4):
(e) the Convention on the Prohibition of the Development, Production and Stockpiling of Bacteriological (Biological) and Toxin Weapons and on their Destruction of 10 April 1972 (the text of which is set out in Schedule 5)"

Thus, although the treaties referred to do not require implementation to include domestic law provisions, the Act claims to do that.

The SPNFZ ?Treaty uses the expression 'nuclear explosive device' which appears in some other treaties concerning nuclear weapons That raises the question of why it was thought necessary by some treaty drafters to introduce the conjunction of nuclear weapons *and/or* explosive nuclear devices.

Rather than binding all people who are New Zealand citizens or

48

ordinarily resident in New Zealand from activities related to nuclear weapons globally, it is only servants of the Crown who are so bound by Article 5:

> **"5 Prohibition on acquisition of nuclear explosive devices**
> (1) No person, who is a New Zealand citizen or a person ordinarily resident in New Zealand, shall, within the New Zealand Nuclear Free Zone,—
> (a) manufacture, acquire, or possess, or have control over, any nuclear explosive device; or
> (b) aid, abet, or procure any person to manufacture, acquire, possess, or have control over any nuclear explosive device.
> (2) No person, who is a New Zealand citizen or a person ordinarily resident in New Zealand, and who is a servant or agent of the Crown, shall, beyond the New Zealand Nuclear Free Zone,—
> (a) manufacture, acquire, or possess, or have control over, any nuclear explosive device; or
> (b) aid, abet, or procure any person to manufacture, acquire, possess, or have control over any nuclear explosive device.

Uniquely, the Act provides the public with a mechanism to hold the government accountable to the legislation through the appointment of an eight member Public Advisory Committee on Disarmament and Arms Control chaired by a Minister of Disarmament and Arms Control. (Clauses 16-19)

Following the 1987 Act, the New Zealand government and civil society activists in New Zealand have played active roles in major initiatives such as the, the World Court Project, Australia Nuclear Warfare Tribunal and subsequently negotiation of the TPNW. These required global civil society coordination and much diplomatic activity at the UN and more generally. More details of the World Court Project, and the Australia Nuclear Warfare Tribunalcan be found in NWIL:3.

5.4 Comparing TPNW and NZ 1987 Act Prohibitions

TPNW: develop, test, produce, manufacture, otherwise acquire, possess or stockpile, assist, encourage or induce, in any way, anyone to engage in any activity prohibited to a State Party under this Treaty in respect of nuclear weapons or other nuclear explosive devices.

NZ 1987 Act: manufacture, acquire, possess, have control over, aid, abet, or procure any person to manufacture, acquire, possess, or have control over any nuclear explosive device.

Applying principles of homonym and synonym detection and resolution when comparing these prohibitions, they are, to all intents and purposes, equivalent. Common mind(s)/drafter(s) somewhere?

Chapter 6

Greater Application of the Nuremberg Principles and ICC

6.1 Introduction

Treaties are agreements between nation states. In the current ultra-statist world, in almost all cases the legal entities involved in negotiating and ratifying or acceding to treaties are nation states. There is no role for individuals in this world of treaties; there are no mechanisms for individuals to accede to treaties. To bind individuals to treaty provisions it is necessary for nation states to enact domestic legislation to bind individual citizens to treaty provisions.

This ultra-statist view of a world of abstract legal entities - nation states - results in a great deal of anthropomorphism in discussions about 'state' behaviour. For long (and it is even attempted today) individuals who commit crimes under international law have attempted to hide their crimes behind their state, denying individual liability.

This tension between state and individual liability came to a head during WWII when the allied powers decided to press ahead after the war, with trials of Axis powers' individuals who were changed in the Nuremberg and Tokyo Tribunals.

The simple reality is that wars are started by individuals acting in concert with other individuals to pick a fight with other states or groups, in the name of the state. Similarly, treaties, conventions, and legislations are created by individuals, although attributed to states or states party. This was put eloquently when talking about UK legislation, by Bennion (2008, p473) thus:

> "A famous conjecture supposed that a million monkeys dancing for a million years on a million typewriters might at random reproduce the plays of Shakespeare. Acts of Parliament are not produced at random, or by monkeys. Neither are they yet produced (as in the future they may conceivably be) by computers. Under our present system Acts are produced, down to the last word and comma, by people. The law maker may be difficult to identify...".

Similarly, the war makers and perpetrators of crimes under international law may be difficult to identify.

TPNW require States Parties to implement the Treaty in their own country (Article 5). Civil society should keep an eye on this and monitor how States Party to TPNW implement their treaty obligations by domestic legislation, regulations, and processes.

International penal law related to nuclear weapons is only developing and there are serious weaknesses that need to be addressed in the coming years. This applies to international penal law generally, and not just in relation to nuclear weapons. We only need to look around us to see the considerable international lack of action against individuals responsible for ongoing serious breaches of the laws of war and international humanitarian law. How many regimes these days are responsible for killing and injuring peaceful non-combatants?

Since WWII there has been an evolution from the Nuremberg Charter, via various UNGA inspired investigations, to a codification of principles of international penal law eventually realized in the Statue of Rome and the ICC. That journey is explored in the following sections. More detail can be found in NWIL3 Chapter 17, on which this chapter draws heavily for its summaries.

Today, international enforcement is very weak. This is another point to be taken on board by civil society in moving forward from the TPNW.

6.2 Nuremberg Charter and Principles

War and armed conflict usually involve considerable cruelty, not only to combatants but also to others not involved in the hostilities; many people involved in armed conflict seize the opportunity for base bestial behaviour and cease showing the restraints usually present in the absence of armed hostilities. Over the centuries prescriptions against cruelty have emerged and some military codes developed. Customary international humanitarian law has emerged to provide important behavioural constraints, but the key problem of enforcement remains.

WWII may well have been when the world witnessed to most extensive barbarism of combatants in warfare. There was a determination by one group of nations to see a codification of international penal law that could be applied to individuals involved is the barbarism of WWII. That codification was formulated by the WWII victors and applied to the defeated. It remains extremely unfortunate that the principles developed during that codification were not applied equally to all combatants. If that had happened, there would also have been trials of decision makers and individuals involved in acts such as the bombings of Hiroshima and Nagasaki, the fire bombings of German and Japanese cities, and other such atrocities being investigated according to fair legal processes. We will never know what the result would have been. It would be interesting to see civil society and international lawyers set up independent tribunals to investigate these matters, as far as would be possible given the passage of time since the events in question.

Although there had been earlier discussions, the first formal start in the emergence of the Nuremberg Principles was the Moscow Declaration in 1943 (Moscow, 1943):

"The United Kingdom, the United States and the Soviet Union have received from many quarters evidence of atrocities, massacres and cold-blooded mass executions which are being perpetrated by the Hitlerite forces in many of the countries they have overrun and from which they are now being steadily expelled. The brutalities of Hitlerite domination are no new thing and all peoples or territories in their grip have suffered from the worst form of Government by terror. What is new is that many of these territories are now being redeemed by the advancing armies of the liberating Powers and that, in their desperation, the recoiling Hitlerite Huns are redoubling their ruthless cruelties. This is now evidenced with particular clearness by the monstrous crimes of the Hitlerites on the territory of the Soviet Union which is being liberated from the Hitlerites, and on French and Italian territory.

Accordingly the aforesaid three Allied Powers, speaking in the interests of the 32 United Nations, hereby solemnly declare and give full warning of their declaration as follows: At the time of the granting of any armistice to any Government which may be set up in Germany, those German officers and men and members of the Nazi party who have been responsible for or have taken a consenting part in the above atrocities, massacres and executions will be sent back to the countries in which their abominable deeds were done in order that they may be judged and punished according to the laws of those liberated countries and of the Free Governments which will be erected therein. Lists will be compiled in all possible detail from all these countries having regard especially to the invaded parts of the Soviet Union, to Poland and Czechoslovakia, to Yugoslavia and Greece including Crete and other islands, to Norway, Denmark, the Netherlands, Belgium, Luxemburg, France and Italy.

Thus Germans who take part in wholesale shootings of Polish officers or in the execution of French, Dutch, Belgian or Norwegian hostages or of Cretan peasants, or who have shared in the slaughters inflicted on the people of Poland or in the territories of the Soviet Union which are now being swept clear of the enemy, will know that they will be brought back to the scene of their crimes and judged on the spot by the peoples whom they have outraged. Let those who have hitherto not imbrued their hands with innocent blood beware lest they join the ranks of the guilty, for most assuredly the three Allied Powers will pursue them to the uttermost ends of the earth and will deliver them to the accusers in order that justice may be done.

The above declaration is without prejudice to the case of German criminals, whose offences have no particular geographical location and who will be punished by a joint decision of the Governments of the Allies."

At the time of the Moscow Declaration, there was already a skeleton

United Nations, referred to in the previous quotation.

The Nuremberg Charter was signed on 8th August 1945, ironically between the atomic bombings of Hiroshima and Nagasaki

The UN Charter entered into force on 24th October 1045.

The Nuremberg trials started on 20th November 1945 with judgment given on 1st October 1946 (Nuremberg, 1946).

The next major event was on 11th December 1946 when the UN General Assembly noted the Charters of the Nuremberg and Tokyo Military Tribunals and affirmed the principles of international law in the Nuremberg Charter and Judgment:

> "*The General Assembly*,
> *Recognizes* the obligation laid upon it by Article 13 , paragraph 1, sub-paragraph a, of the Charter, to initiate studies and make recommendations for the purpose of encouraging the progressive development of international law and its codification;
> *Takes note* of the Agreement for the establishment of an International Military Tribunal for the prosecution and punishment of the major war criminals of the European Axis signed in London on 8 August 1945, and of the Charter annexed thereto, and of the fact that similar principles have been adopted in the Charter of the International Military Tribunal for the trial of the major war criminals in the Far East, proclaimed at Tokyo on 19 January 1946;
> *Therefore*,
> *Affirms* the principles of international law recognized by the Charter of the Niirnberg Tribunal and the judgment of the Tribunal;
> *Directs* the Committee on the codification of international law established by the resolution of the General Assembly of 11 December 1946, to treat as a matter of primary importance plans for the formulation, in the context of a general codification of offences against the peace and security of mankind, or of an International Criminal Code, of the principles recognized in the Charter of the Niirnberg Tribunal and in the judgment of the Tribunal.
> Fifty-fifth plenary meeting, 11 December 1946." (UNGA/95, 1946)

On 21st November 1947 the UN General Assembly set up the International Law Commission and remitted to it the task of codifying principles of international law based on the Nuremberg Charter and judgment (UNGA/177, 1947).

The Nuremberg Principles set out in Chapter 3 are the Principles adopted in the 2nd session of the International Law Commission.

The International Law Commission put in considerable work in codifying the Nuremberg Charter and Judgment. Detailed summaries of the work are contained in Volume II of the 1950 Yearbook of the International Law Commission (ILC, 1950a).

The crimes that are included in the draft code are: crimes against peace, war crimes, and crimes against humanity. Aggressive war was left out due to difficulties in defining 'aggression'. Detailed discussion can be found in the 1950 ILC Yearbook, vol II, in the section about the formulation of Nuremberg Principles (ILC, 1950b). The discussion underpinning Principle VII (Complicity) has more detail in the ILC Yearbook about a Draft Code of Offences Against the Peace and Security of Mankind (ILC, 1950c). That sets out a set of crimes as:

"The parties to the code declare that the acts mentioned below are crimes under international law which they undertake to prevent and punish :

Crime No. I   The use of armed force in violation of international law and, in particular, the waging of aggressive war.

Crime No. II   The invasion by armed gangs of the territory of another State.

Crime No. III   The fomenting, by whatever means, of civil strife in another State.

Crime, No. IV   Organized terroristic activities carried out in another State.

Crime No. V   Manufacture, trafficking and possession of weapons the use of which is prohibited by international agreements.

Crime No. VI   The violation of military clauses of international treaties defining the war potential of a State, namely, clauses concerning;

(a) The strength of land, sea and air forces.

(b) Armaments, munitions and war material in general.

(c) Presence of land, sea and air forces, armaments, munitions and war material.

(d) Recruiting and military training.

(e) Fortifications.

Crime No. VII   The annexation of territories in violation of international law.

Crime No. VIII

1. The Commission of any the following acts committed with intent to destroy, in whole or in part, a national, ethnical, racial or religious group, as such:

(a) Killing members of the group;

(b) Causing serious bodily or mental harm to members of the group;

(c) Deliberately inflicting on the group conditions of life calculated to bring about its physical destruction in whole or in part;

(d) Imposing measures intended to prevent births within the group;

(e) Forcibly transferring children of the group to another group.

2. The commission of any of the following acts in so far as they are not covered by the foregoing paragraph:

Murder, extermination, enslavement, deportation and other inhuman acts done against a civilian population, or persecutions on political, racial or religious grounds when such acts are done or such persecutions are carried on in execution of or in connexion with any crime against peace or war crimes as defined by the Charter of the International Military Tribunal.

Crime No. IX   Violations of the laws or customs of war.[namely,. . .]

Crime No. X

(a) Conspiracy to commit any of the acts enumerated under Crimes I-IX.

(b) Direct and public incitement to commit any of the acts under Crimes I-IX.

(c) Preparatory acts to commit any of the acts under Crimes I-IX.

(d) Attempt to commit any of the acts under Crimes I-IX.

(e) Complicity in any of the acts under Crimes I-IX" (ILC 1950c; pp 277-278).

The list above can be referred to as the ILC 1950 List of Offences Against the Peace and Security of Mankind (1950 LOAPSM).

The ILC submitted its Report containing the draft Nuremberg Principles to UNGA. UNGA noted the Report but did not formally vote to accept the Principles.

6.3 Rome Statute and the International Criminal Court (ICC)

In 1950,the ILC changed focus, as did UNGA. In 1947 UNGA gave the task of codifying Nuremberg Principles out of the Nuremberg Charter and Judgment. ILC finished that task in 1950 and reported back to UNGA. UNGA noted the work in the Report. Job done.

As can be seen at the end of the previous section, UNGA started to work on Offences Against the Peace and Security of Mankind.

The draft code of crimes evolved over the following 40+ years emerging in a much more developed form (ILC, 1996) and finally emerged with modifications in the Rome Statute (ICC 2011). The discussion around the formulation of this code already contains the beginning of a need for an international court to cover individual actions and responsibilities, which eventually emerged as the International Criminal Court.

In its deliberations in 1996 the ILC discussed adding "wilful and severe damage to the environment be considered either as a war crime, or a crime against humanity, or a separate crime against the peace and security of mankind" (ILC, 1996; p43).

The International Law Commission produced a full draft code of Crimes Against the Peace and Security of Mankind in 1996 (ILC, 1996). That code includes damage to the environment due to military action (Article 20(g)). That draft code eventually emerged revised and codified in the Rome Statute on 17th July 1998 (ICC, 1998) setting up the International Criminal Court (ICC) which came into force on 1st July 2002. There have been some amendments to the Rome Statute since entering into force, and the current version is available from the ICC website (ICC, 2011). It is worth noting that the Rome Statute includes damage to the environment; it is not exclusively concerned with impact on humans. The ICC is focussed on individual liability.

The following subsections give the key sections of the Rome Statute that identify the main heads of international penal law.

Article 5 Crimes within the jurisdiction of the Court

The jurisdiction of the Court shall be limited to the most serious crimes of concern to the international community as a whole. The Court has jurisdiction in accordance with this Statute with respect to the following crimes:
    (a) The crime of genocide;
    (b) Crimes against humanity;
    (c) War crimes;
    (d) The crime of aggression.

Article 6 Genocide

For the purpose of this Statute, "genocide" means any of the following acts committed with intent to destroy, in whole or in part, a national, ethnical, racial or religious group, as such:
    (a) Killing members of the group;
    (b) Causing serious bodily or mental harm to members of the group;
    (c) Deliberately inflicting on the group conditions of life calculated to bring about its physical destruction in whole or in part;
    (d) Imposing measures intended to prevent births within the group;
    (e) Forcibly transferring children of the group to another group.

Article 7 Crimes against humanity

1. For the purpose of this Statute, "crime against humanity" means any of the following acts when committed as part of a widespread or systematic attack directed against any civilian population, with knowledge of the attack:
    (a) Murder;
    (b) Extermination;
    (c) Enslavement;
    (d) Deportation or forcible transfer of population;
    (e) Imprisonment or other severe deprivation of physical liberty in violation of fundamental rules of international law;
    (f) Torture;
    (g) Rape, sexual slavery, enforced prostitution, forced pregnancy, enforced sterilization, or any other form of sexual violence of comparable gravity;
    (h) Persecution against any identifiable group or collectivity on political, racial, national, ethnic, cultural, religious, gender as defined in paragraph 3, or other grounds that are universally recognized as impermissible under international law, in connection with any act referred to in this paragraph or any crime within the jurisdiction of the Court;
    (i) Enforced disappearance of persons;
    (j) The crime of apartheid;

(k) Other inhumane acts of a similar character intentionally causing great suffering, or serious injury to body or to mental or physical health.

2. For the purpose of paragraph 1:

(a) "Attack directed against any civilian population" means a course of conduct involving the multiple commission of acts referred to in paragraph 1 against any civilian population, pursuant to or in furtherance of a State or organizational policy to commit such attack;

(b) "Extermination" includes the intentional infliction of conditions of life, inter alia the deprivation of access to food and medicine, calculated to bring about the destruction of part of a population;

(c) "Enslavement" means the exercise of any or all of the powers attaching to the right of ownership over a person and includes the exercise of such power in the course of trafficking in persons, in particular women and children;

(d) "Deportation or forcible transfer of population" means forced displacement of the persons concerned by expulsion or other coercive acts from the area in which they are lawfully present, without grounds permitted under international law;

(e) "Torture" means the intentional infliction of severe pain or suffering, whether physical or mental, upon a person in the custody or under the control of the accused; except that torture shall not include pain or suffering arising only from, inherent in or incidental to, lawful sanctions;

(f) "Forced pregnancy" means the unlawful confinement of a woman forcibly made pregnant, with the intent of affecting the ethnic composition of any population or carrying out other grave violations of international law. This definition shall not in any way be interpreted as affecting national laws relating to pregnancy;

(g) "Persecution" means the intentional and severe deprivation of fundamental rights contrary to international law by reason of the identity of the group or collectivity;

(h) "The crime of apartheid" means inhumane acts of a character similar to those referred to in paragraph 1, committed in the context of an institutionalized regime of systematic oppression and domination by one racial group over any other racial group or groups and committed with the intention of maintaining that regime;

(i) "Enforced disappearance of persons" means the arrest, detention or abduction of persons by, or with the authorization, support or acquiescence of, a State or a political organization, followed by a refusal to acknowledge that deprivation of freedom or to give information on the fate or whereabouts of those persons, with the intention of removing them from the protection of the law for a prolonged period of time.

3. For the purpose of this Statute, it is understood that the term "gender" refers to the two sexes, male and female, within the context of society. The term "gender" does not indicate any meaning different from the above.

Article 8 War crimes

1. The Court shall have jurisdiction in respect of war crimes in particular when committed as part of a plan or policy or as part of a large-scale commission of such crimes.

2. For the purpose of this Statute, "war crimes" means:

   (a) Grave breaches of the Geneva Conventions of 12 August 1949, namely, any of the following acts against persons or property protected under the provisions of the relevant Geneva Convention:

   (i) Wilful killing;

   (ii) Torture or inhuman treatment, including biological experiments;

   (iii) Wilfully causing great suffering, or serious injury to body or health;

   (iv) Extensive destruction and appropriation of property, not justified by military necessity and carried out unlawfully and wantonly;

   (v) Compelling a prisoner of war or other protected person to serve in the forces of a hostile Power;

   (vi) Wilfully depriving a prisoner of war or other protected person of the rights of fair and regular trial;

   (vii) Unlawful deportation or transfer or unlawful confinement;

   (viii) Taking of hostages.

   (b) Other serious violations of the laws and customs applicable in international armed conflict, within the established framework of international law, namely, any of the following acts:

   (i) Intentionally directing attacks against the civilian population as such or against individual civilians not taking direct part in hostilities;

   (ii) Intentionally directing attacks against civilian objects, that is, objects which are not military objectives;

   (iii) Intentionally directing attacks against personnel, installations, material, units or vehicles involved in a humanitarian assistance or peacekeeping mission in accordance with the Charter of the United Nations, as long as they are entitled to the protection given to civilians or civilian objects under the international law of armed conflict;

   (iv) Intentionally launching an attack in the knowledge that such attack will cause incidental loss of life or injury to civilians or damage to civilian objects or widespread, long-term and severe damage to the natural environment which would be clearly excessive in relation to the concrete and direct overall military advantage anticipated;

   (v) Attacking or bombarding, by whatever means, towns, villages, dwellings or buildings which are undefended and which are not military objectives;

   (vi) Killing or wounding a combatant who, having laid down his arms or having no longer means of defence, has surrendered at discretion;

(vii) Making improper use of a flag of truce, of the flag or of the military insignia and uniform of the enemy or of the United Nations, as well as of the distinctive emblems of the Geneva Conventions, resulting in death or serious personal injury;

(viii) The transfer, directly or indirectly, by the Occupying Power of parts of its own civilian population into the territory it occupies, or the deportation or transfer of all or parts of the population of the occupied territory within or outside this territory;

(ix) Intentionally directing attacks against buildings dedicated to religion, education, art, science or charitable purposes, historic monuments, hospitals and places where the sick and wounded are collected, provided they are not military objectives;

(x) Subjecting persons who are in the power of an adverse party to physical mutilation or to medical or scientific experiments of any kind which are neither justified by the medical, dental or hospital treatment of the person concerned nor carried out in his or her interest, and which cause death to or seriously endanger the health of such person or persons;

(xi) Killing or wounding treacherously individuals belonging to the hostile nation or army;

(xii) Declaring that no quarter will be given;

(xiii) Destroying or seizing the enemy's property unless such destruction or seizure be imperatively demanded by the necessities of war;

(xiv) Declaring abolished, suspended or inadmissible in a court of law the rights and actions of the nationals of the hostile party;

(xv) Compelling the nationals of the hostile party to take part in the operations of war directed against their own country, even if they were in the belligerent's service before the commencement of the war;

(xvi) Pillaging a town or place, even when taken by assault;

(xvii) Employing poison or poisoned weapons;

(xviii) Employing asphyxiating, poisonous or other gases, and all analogous liquids, materials or devices;

(xix) Employing bullets which expand or flatten easily in the human body, such as bullets with a hard envelope which does not entirely cover the core or is pierced with incisions;

(xx) Employing weapons, projectiles and material and methods of warfare which are of a nature to cause superfluous injury or unnecessary suffering or which are inherently indiscriminate in violation of the international law of armed conflict, provided that such weapons, projectiles and material and methods of warfare are the subject of a comprehensive prohibition and are included in an annex to this Statute, by an amendment in accordance with the relevant provisions set forth in articles 121 and 123;

(xxi) Committing outrages upon personal dignity, in particular humiliating and degrading treatment;

(xxii) Committing rape, sexual slavery, enforced prostitution, forced pregnancy, as defined in article 7, paragraph 2 (f), enforced sterilization, or any other form of sexual violence also constituting a grave breach of the Geneva Conventions;

(xxiii) Utilizing the presence of a civilian or other protected person to render certain points, areas or military forces immune from military operations;

(xxiv) Intentionally directing attacks against buildings, material, medical units and transport, and personnel using the distinctive emblems of the Geneva Conventions in conformity with international law;

(xxv) Intentionally using starvation of civilians as a method of warfare by depriving them of objects indispensable to their survival, including wilfully impeding relief supplies as provided for under the Geneva Conventions;

(xxvi) Conscripting or enlisting children under the age of fifteen years into the national armed forces or using them to participate actively in hostilities.

(c) In the case of an armed conflict not of an international character, serious violations of article 3 common to the four Geneva Conventions of 12 August 1949, namely, any of the following acts committed against persons taking no active part in the hostilities, including members of armed forces who have laid down their arms and those placed hors de combat by sickness, wounds, detention or any other cause:

(i) Violence to life and person, in particular murder of all kinds, mutilation, cruel treatment and torture;

(ii) Committing outrages upon personal dignity, in particular humiliating and degrading treatment;

(iii) Taking of hostages;

(iv) The passing of sentences and the carrying out of executions without previous judgement pronounced by a regularly constituted court, affording all judicial guarantees which are generally recognized as indispensable.

(d) Paragraph 2 (c) applies to armed conflicts not of an international character and thus does not apply to situations of internal disturbances and tensions, such as riots, isolated and sporadic acts of violence or other acts of a similar nature.

(e) Other serious violations of the laws and customs applicable in armed conflicts not of an international character, within the established framework of international law, namely, any of the following acts:

(i) Intentionally directing attacks against the civilian population as such or against individual civilians not taking direct part in hostilities;

(ii) Intentionally directing attacks against buildings, material, medical units and transport, and personnel using the distinctive emblems of the Geneva Conventions in conformity with international law;

(iii) Intentionally directing attacks against personnel, installations, material, units or vehicles involved in a humanitarian assistance or peacekeeping mission in accordance with the Charter of the United Nations, as long as they are entitled to the protection given to civilians or civilian objects under the international law of armed conflict;

(iv) Intentionally directing attacks against buildings dedicated to religion, education, art, science or charitable purposes, historic monuments, hospitals and places where the sick and wounded are collected, provided they are not military objectives;

(v) Pillaging a town or place, even when taken by assault;

(vi) Committing rape, sexual slavery, enforced prostitution, forced pregnancy, as defined in article 7, paragraph 2(f), enforced sterilization, and any other form of sexual violence also constituting a serious violation of article 3 common to the four Geneva Conventions;

(vii) Conscripting or enlisting children under the age of fifteen years into armed forces or groups or using them to participate actively in hostilities;

(viii) Ordering the displacement of the civilian population for reasons related to the conflict, unless the security of the civilians involved or imperative military reasons so demand;

(ix) Killing or wounding treacherously a combatant adversary;

(x) Declaring that no quarter will be given;

(xi) Subjecting persons who are in the power of another party to the conflict to physical mutilation or to medical or scientific experiments of any kind which are neither justified by the medical, dental or hospital treatment of the person concerned nor carried out in his or her interest, and which cause death to or seriously endanger the health of such person or persons;

(xii) Destroying or seizing the property of an adversary unless such destruction or seizure be imperatively demanded by the necessities of the conflict;

(xiii) Employing poison or poisoned weapons;

(xiv) Employing asphyxiating, poisonous or other gases, and all analogous liquids, materials or devices;

(xv) Employing bullets which expand or flatten easily in the human body, such as bullets with a hard envelope which does not entirely cover the core or is pierced with incisions.

(f) Paragraph 2 (e) applies to armed conflicts not of an international character and thus does not apply to situations of internal disturbances and tensions, such as riots, isolated and sporadic acts of violence or other acts of a similar nature. It applies to armed conflicts that take place in the territory of a State when there is protracted armed conflict between governmental authorities and organized armed groups or between such groups.3. Nothing in paragraph 2 (c) and (e) shall affect the responsibility of a Government to maintain or re-establish law and order in the State or to defend the unity and territorial integrity of the State, by all legitimate means.

Article 8 bis[1] Crime of aggression

1. For the purpose of this Statute, "crime of aggression" means the planning, preparation, initiation or execution, by a person in a position effectively to exercise control over or to direct the political or military action of a State, of an act of aggression which, by its character, gravity and scale, constitutes a manifest violation of the Charter of the United Nations.
2. For the purpose of paragraph 1, "act of aggression" means the use of armed force by a State against the sovereignty, territorial integrity or political independence of another State, or in any other manner inconsistent with the Charter of the United Nations. Any of the following acts, regardless of a declaration of war, shall, in accordance with United Nations General Assembly resolution 3314 (XXIX) of 14 December 1974, qualify as an act of aggression:

(a) The invasion or attack by the armed forces of a State of the territory of another State, or any military occupation, however temporary, resulting from such invasion or attack, or any annexation by the use of force of the territory of another State or part thereof;

(b) Bombardment by the armed forces of a State against the territory of another State or the use of any weapons by a State against the territory of another State;

(c) The blockade of the ports or coasts of a State by the armed forces of another State;

(d) An attack by the armed forces of a State on the land, sea or air forces, or marine and air fleets of another State;

6.4 International Criminal Court (ICC) and taking TPNW forward

Although it can be demonstrated that the Rome Statute origins lie in the Nuremberg Charter and Judgment, and the 50+ year journey via the ILC and UNGA, there are some fundamental differences that mean it is necessary to complement the Rome Charter with elements of the Nuremberg Charter and Principles.

1 Inserted by resolution RC/Res.6 of 11 June 2010

The Nuremberg Charter and Principles were concerned with Crimes against Peace, War Crimes and Crimes against Humanity. War of aggression (aggressive war) is included as part of Crimes against Peace.

As noted above, after the ILC finished its codification of the Nuremberg Charter and Judgment into the Nuremberg Principles, it devoted its time to considering Offences Against the Peace and Security of Mankind. The first draft of their code is shown above.

A fundamental weakness in the Rome Stature is that when it was formulated, it lost the 'Crimes against Peace' from Nuremberg. The early problems the ILC had over defining 'aggression' were solved eventually by UNGA in 1974, (UNGA/3314, 1974) as discussed above and in more detail in NWIL3.

Article 6 of the Nuremberg Charter provides that:
"...

...

The following acts, or any of them, are crimes coming within the juris diction of the Tribunal for which there shall be individual responsibility:-

(a) Crimes against peace: namely, planning, preparation, initiation or waging of a war of aggression, or a war in violation of international treaties, agreements or assurances, or participation in a common plan or conspiracy for the accomplishment of any of the foregoing;

...

..." (Nuremberg, 1945)

As the Rome Statute is concerned with war crimes, crimes against humanity, and/or aggression, it becomes relevant *after* any use of nuclear weapons. It is clear that any individuals involved in any use of nuclear weapons would be indictable under all three headings — war crimes, crimes against humanity, and aggression. Tje Crime of Aggression is available *before* any actual use of nuclear weapons.

Any individuals involved in nuclear weapons *before* any use are indictable under the Nuremberg Charter or Principles under the heads of Crimes against Peace, and Complicity.

Civil society can do some things:

1. Support creating and maintaining a Nuclear Weapons Register of individuals, corporations, and organizations involved with nuclear weapons;

2. Campaign for an international mechanism to deal with Crimes against Peace;

3. Set up tribunals to hear cases of Crimes against Peace.

Chapter 7

Case Study: UK and Trident

7.1 Introduction

Since before the atomic bombings of Hiroshima and Nagasaki, the UK and USA have been working together to produce nuclear weapons and their delivery systems.

The term 'atomic bomb' had emerged around 1938 in discussions among scientists who had escaped from Hitler's Germany during the 1930s and were working with a number of British scientists.

The discussions, collaborations, and conferences in which nuclear scientists were involved in the late 1930s and early 1940s reached politicians in the UK and USA who realised that Germany was also involved in efforts to build a 'super-bomb'. The UK and USA governments decided to establish the Manhattan Project staffed with USA and UK scientists in conjunction with various scientists who had escaped from Nazi Germany. The British mission to the Manhattan Project was established in Los Alamos in 1943. The UK contribution to the Manhattan Project was crucial for its successful development of atomic weapons.

For an interesting history of the UK involvement in Manhattan see Szasz (1992) who observes:

> "During World War 2, President Franklin D. Roosevelt and Prime Minister Winston Churchill pooled their nations' resources in the desperate race to beat the Germans to the secret of the atomic bomb." (Szasz, 1992; inside front cover).

It is beyond the scope of this book to go into the early history of the atomic bomb. Suffice it to say that nuclear weapons have emerged and developed within a context of a close relationship between the UK and USA which goes right back to the original birth of the bomb.

Since those early days of the 1930s and 1940s, the UK and USA have moved apart in terms of the development of nuclear weapons. Nonetheless, the relationship between the two remains very strong particularly with respect to delivery systems, of which Trident is the current example.

It may well be that cooperation over nuclear weapons and explosive devices between the UK and USA has now entered the world of state secrets (and could fall foul of the NPT). Rumours continue that since the significant reduction in nuclear power generation in the USA due to a privatized system of reactors has exposed the commercial folly of nuclear power generation, the UK has been meeting part of the US nuclear weapons fuel gap, hence an increased UK government promotion of nuclear power generation. Of course, it is not only nuclear weapons that require nuclear fuels, demand has also increased because of an increasing use of nuclear propulsion in military vehicles such as aircraft carriers and submarines.

It is astonishing how the drafters of treaties such as the NPT limit considerations to nuclear weapons and explosive nuclear devices, thus leaving out delivery vehicles. There are also many clauses permitting the 'peaceful' uses of nuclear power thereby permitting international trade in nuclear fuels that are intended for use in nuclear weapons.

So the relationship between the USA and UK concerning nuclear weapons started with a very close scientific collaboration between them for the development and production of the first nuclear weapons. There was some drifting apart in a context of Soviet espionage and the UK decided to develop its own 'independent' nuclear weapons. That was long before the NPT. Of course, today we don't know about 'secret' arrangements between the countries, but we do know about the 1958 US-UK Mutual Defence Agreement (see UK-US, 1958) which is updated and renewed from time to time. The matter of the intimate relationship between nuclear reactors, power generation, and nuclear weapons requires its own book.

During the drafting of this book, the Prime Minister at the time, Boris Johnson, announced to the House of Commons a 40% increase of the cap on the number of nuclear weapons (Mills, 2021), accompanied by commentary that this will involve the UK breaching its obligations under the NPT article VI and encourage nuclear weapons proliferation (Stone, 2021).

The rest of this chapter looks at a summary of the UK's nuclear weapons policies, views from the front line of Trident deployment, and the 2021 UK government decision to increase the cap on nuclear warheads.

7.2 Development of UK Nuclear Weapons Policy

This section is a summary of material in NWIL3 which has two parts of particular relevance: Appendix C, which is the Application by the Republic of the Marshall Islands (RMI) in its case against the UK in the ICJ for breaches of the UK obligations under the NPT; and Chapter 13 which is Nick Grief's commentary on the ICJ judgments in the RMI v UK case (and related cases).

In order to keep this material as a summary, all the original footnotes have been removed, but they are readily available in NWIL3. Paragraph numbering has been retained.

C. *The UK and the Nuclear Arms Race*

1. Early Nuclear History

24. On 3 October 1952, the first British atomic device was detonated in the Monte Bello Islands off north-western Australia. On 7 November 1953, the UK's first operational atomic bomb, the Blue Danube, arrived at RAF Wittering from AWE Aldermaston.

25. On 26 July 1954 the Cabinet agreed to the manufacture of a much more powerful British hydrogen bomb and on 15 May 1957 the UK tested a thermonuclear device at Christmas Island in the Pacific.

26. On 4 August 1958, the U.S. and UK governments concluded the Agreement for Co-operation on the Uses of Atomic Energy for Mutual Defence Purposes (the "Mutual Defence Agreement" or "MDA").

27. On 3 January 1963, the Cabinet authorized the purchase of Polaris C3 submarine-launched ballistic missiles and re-entry vehicles from the U.S. Government. On January 25, 1965, the decision was taken to build four Resolution-class submarines to carry the Polaris missiles, partly to ensure that one boat would always be on station when the Royal Navy assumed the main nuclear weapons system role in the late 1960s. HMS Resolution, the first of the four Polaris missile-carrying submarines, was commissioned on 30 October 1967 and on June 14, 1968, Polaris submarines formally took over the primary strategic nuclear weapons deployment role from the RAF's 'V' bomber force.

28. The development of the Super Antelope (later known as Chevaline) re-entry body for the UK's Polaris warheads was approved on 30 October 1973. This was because the UK could no longer be certain that a sufficient number of Polaris warheads would penetrate Soviet ABM defences to cause the damage required to exert a credible deterrent effect. In November 1982 the Ministry of Defence announced

that Chevaline-equipped missiles were operational at sea.

29. In July 1980, the UK government announced the decision to buy the U.S. Trident C4 missile system as a replacement for the Polaris system, which was due to reach the end of its service life in the early 1990s. In March 1982, however, the order was changed to the Trident II D5, a new missile announced by the U.S. in October 1981. This ensured missile commonality between the U.S. Navy and the Royal Navy. The UK defence establishment wanted to ensure that any future UK nuclear system remained in step with U.S. nuclear hardware and weapon programmes after the difficult experience with the indigenous Chevaline upgrade. Former Permanent Under Secretary at the Ministry of Defence, Sir Michael Quinlan, stated in 2004 that "Purely in weight of strike potential the United Kingdom could have been content with less than Trident could offer, even in the C4 version originally chosen (let alone D5 version to which the United Kingdom switched in early 1982, when it had become clear that the United States was committed to proceed with its acquisition and deployment). The original choice and the switch were driven in large measure by the long-term financial and logistic benefits of commonality with the United States".

2. The UK's Current Nuclear Arsenal

30. The UK's nuclear weapons system is based upon the submarine-launched Trident D5 missile. It is the UK's third-generation strategic nuclear weapon system. Trident was procured during the final decade of the Cold War and was brought into service to replace Polaris over a six-year period beginning in December 1994. It is now the UK's only nuclear weapons system, the UK having retired its air-launched WE177 free-fall nuclear bombs and repatriated forward-deployed US tactical nuclear weapons operated by UK forces under dual-key arrangements in the 1990s.

31. The Trident nuclear weapons system has three technical components.

a) The Vanguard-class nuclear-powered ballistic submarines (SSBN), of which the UK has four: HMS Vanguard, HMS Victorious, HMS Vigilant and HMS Vengeance, designed and built in the UK by Vickers Shipbuilding and Engineering Ltd (VSEL), now BAE Systems, in Barrow-in-Furness, Cumbria. Refit and maintenance are carried out by Devonport Management Limited in Devonport, Plymouth, UK.

b) The Trident D5 submarine-launched intercontinental ballistic missile (ICBM), manufactured in the US by Lockheed Martin. Under the Polaris Sales Agreement as modified for

Trident, the UK has title to 58 missiles. Aside from those currently deployed, the missiles are held in a communal pool at the US Strategic Weapons facility at King's Bay, Georgia, US. Each submarine is capable of carrying up to 16 Trident D5 missiles.

c) The components for the nuclear warheads, including qualitative improvements to them, are made in the UK at the Atomic Weapons Establishment (AWE) Aldermaston, Berkshire, and assembled at nearby AWE Burghfield, Berkshire. There is extensive collaboration between the UK and the US on the production of the UK 's warheads under the Mutual Defence Agreement, "which provides for extensive cooperation on nuclear warhead and reactor technologies, in particular the exchange of classified information concerning nuclear weapons to improve 'design, development and fabrication' capability and the transfer of nuclear warhead-related materials"As a result, some components of the UK warheads are manufactured, and undergo qualitative improvements, in the U.S.

32. The submarine fleet is supported by an extensive onshore infrastructure. The Vanguard submarines are based at HM Naval Base Clyde, Faslane, Scotland. Nuclear warheads are fitted to the D5 missiles at the Royal Naval Armaments Depot Coulport (part of HM Naval Base Clyde). The warheads are transported by road from AWE Burghfield to Coulport, where they are placed in underground bunkers in the Trident Area. When required they are taken to the Explosive Handling Jetty where they are fitted onto the missiles on the Trident submarines.

33. The Strategic Defence Review, published on 8 July 1998, affirmed the Government's commitment to maintaining a nuclear weapons system but made a number of changes to it. The warhead stockpile was to be cut from the ceiling of up to 300 warheads maintained by the previous government to fewer than 200 operationally available warheads. The patrol cycle of the Trident submarines was also relaxed with normally only one submarine on patrol at any one time. As with pre-Chevaline Polaris, each submarine would now carry a maximum of 48 warheads, rather than the ceiling of up to 96. The Trident submarine's alert status was also to be reduced. Missiles had not been targeted for some years but, in addition, submarines would normally now be at several days' rather than 15 minutes' notice to fire. A requirement for an additional seven Trident missile bodies was cancelled, leaving a new total of 58.

34. The Strategic Defence and Security Review, published on 19 October 2010, reaffirmed the UK's commitment to a submarine-launched nuclear weapons system on continuous alert based on the Trident missile delivery system, and announced that: the number of warheads on board each deployed submarine would be reduced from 48 to 40; the requirement for operationally available warheads would be reduced from fewer than 160 to no more than 120; the number of operational missiles on the Vanguard class submarines would be reduced to no more than 8; and the UK's overall nuclear weapons stockpile would be reduced from not more than 225 to no more than 180 by the mid-2020s.

3. Nuclear Policy, Doctrine and Expenditure

35. The Royal Navy has maintained unbroken nuclear weapon patrols since 1968. The 1998 Strategic Defence Review stated that the UK would continue to maintain these continuous-at-sea nuclear armed patrols. This means that one of the four Vanguard-class submarines is on patrol at any given time.

36. Trident is the UK's most advanced nuclear weapon system to date. With a range of between 6,500 kilometres and 12,000 kilometres, depending on payload, Trident's greater speed, accuracy and multiple independently targetable warheads distinguish it from, and enable it to reach more targets than, its predecessor, Polaris Chevaline.

37. As the Defence Select Committee noted in 1994:

> Trident's accuracy and sophistication in other respects does - and was always intended to - represent a significant enhancement of the UK's nuclear capability. We have invested a great deal of money to make it possible to attack more targets with greater effectiveness using nominally equivalent explosive power.

38. Trident was originally designed as a strategic nuclear system with respect to threats posed by the Soviet Union. In 1993, however, following the end of the Cold War, the then Secretary of State for Defence announced that in future Trident's role would be to deter "potential aggressors" from threatening UK "vital interests". In order to do this, Trident was assigned an additional "sub-strategic" role:

> The ability to undertake a massive strike with strategic systems is not enough to ensure deterrence. An aggressor might, in certain circumstances, gamble on a lack of will ultimately to resort to such dire action. It is therefore important for the credibility of our deterrent that the United Kingdom also possesses the capability . to undertake a more limited nuclear strike in order to induce a political decision to halt aggression by delivering an unmistakable message of our willingness to defend our vital interests to the utmost.

39. As part of the agreement under which the UK procured Polaris

and subsequently Trident missiles from the US, UK Trident forces are assigned to NATO to be used for the defence of the Alliance "except where the UK government may decide that supreme national interests are at stake". The UK is therefore committed to NATO's nuclear policy, which since the mid-1960s has been based on a doctrine of "flexible response". One of the key elements of NATO's nuclear doctrine is that the Alliance refuses to rule out the first use of NATO nuclear weapons, thereby allowing its nuclear planners to prepare for that option.

40. Similarly, the UK has always refused to rule out the first use of its nuclear weapons, especially in cases where biological or chemical weapons may have been used. For example, shortly after the 1997 general election, the then Minister of State Dr John Reid stated:

> The role of deterrence ...must not be overlooked. Even if a potential aggressor has developed missiles with the range to strike at the United Kingdom, and nuclear, biological or chemical warheads to be delivered by those means, he would have to consider - he would do well to consider - the possible consequences of such an attack ...It seems unlikely that a dictator who was willing to strike another country with weapons of mass destruction would be so trusting as to feel entirely sure that that country would not respond with the power at its disposal.

41. Following the terrorist attacks on the U.S. in September 2001, a new chapter of the Strategic Defence Review extended the role of nuclear weapons further to include allegedly deterring terrorist organisations:

> The UK's nuclear weapons have a continuing use as a means of deterring major strategic military threats, and they have a continuing role in guaranteeing the ultimate security of the UK. But we also want it to be clear, particularly to the leaders of states of concern and terrorist organisations, that all our forces play a part in deterrence, and that we have a broad range of responses available

42. The implication is that the UK is willing, if deemed to be necessary, to use its nuclear weapons against States of concern[1] and terrorist organisations.

43. The 2010 Strategic Defence and Security Review states that the UK "would only consider using nuclear weapons in extreme circumstances of self-defence, including the defence of our NATO allies", adding: "we remain deliberately ambiguous about precisely when, how and at what scale we would contemplate their use".

44. The Strategic Defence and Security Review reaffirms in modified form existing assurances to non-nuclear-weapon States Parties to the NPT. It states "that the UK will not use or threaten to use nuclear weapons against non-nuclear weapon States parties to the NPT" but notes that "this assurance would not apply to any State in

---

1 Editorial nore: In 2002, the UK Defence Secretary confirmed UK willingness to use nuclear weapons in Iraq in certain circumstances, see Hoon (2002).

material breach of those non-proliferation obligations". It also notes that "while there is currently no direct threat to the UK or its vital interests from States developing capabilities in other weapons of mass destruction, for example chemical and biological, we reserve the right to review this assurance if the future threat, development and proliferation of these weapons make it necessary".

45. The UK has continued to maintain and modernize its nuclear forces with annual expenditure on capital and running costs at around 5 to 6 per cent of the UK defence budget. This does not include costs for recapitalising the Trident system estimated to be £25 billion at outturn prices

4. Current Plans for Modernization and Qualitative Improvements of the UK's Nuclear Arsenal

46. In December 2006 the UK Government published a White Paper which formally opened the process to replace the UK's Trident nuclear weapons system. The White Paper was endorsed by the House of Commons on 14 March 2007 when the following motion was carried by 409 votes to 161:

> That this House supports the Government's decisions, as set out in the White Paper The Future of the United Kingdom's Nuclear Deterrent (Cm 6994), to take the steps necessary to maintain the UK's minimum strategic nuclear deterrent beyond the life of the existing system and to take further steps towards meeting the UK's disarmament responsibilities under Article VI of the Non-Proliferation Treaty.

47. According to British Pugwash, the effect of that vote and its present and future consequences are as follows:

> Parliament voted to authorize the initial 'Concept' phase of the Trident replacement system. The next major milestone, known as the 'Initial Gate' decision, was to move to the 'Assessment' phase, involving further detailed refinement of a set of design options to enable selection of a preferred solution. The government announced the Initial Gate decision on May 18, 2011. The next big decision to move to the 'Demonstration and Manufacture' phase is the 'Main Gate' decision, now scheduled for 2016 (delayed from 2014 in October 2010). That is supposed to be the key decision-point when the finalized submarine design is adopted; contracts to build the new boats are then tendered, and billions more pounds will be irrevocably committed to construction of a new generation of nuclear weapons.

48. The Strategic Defence and Security Review 2010 states:

> Under the 1958 UK-US Agreement for Cooperation on the Uses of Atomic Energy for Mutual Defence Purposes (the 'Mutual Defence Agreement') we have agreed on the future of the Trident D5 delivery system and determined that a replacement warhead is not required until at least the late 2030s.

Decisions on replacing the warhead will not therefore be required in this Parliament. This will defer £500 million of spending from the next 10 years.

49. Under the UK-US Mutual Defence Agreement, a new "arming, fusing and firing system" developed by the US is to be used in current UK warheads. The system would improve the nuclear warhead's effectiveness against hardened targets. The Trident II D5 missile can carry two types of re-entry vehicle (RV) that house each nuclear warhead: the Mark 4 for the U.S. W76 warhead and the Mark 5 for the more modern and higher yield W88 warhead. The UK purchased the Mark 4 RV and designed a warhead to meet Mk4 RV specifications in terms of weight, size, shape, centre of gravity, and centre of inertia. The U.S. is modernizing its W76 warheads and Mk4 re-entry vehicles, including launcher, navigation, fire control, guidance, and re-entry systems. The modernized W76-1 and Mk4A RV have improved the accuracy of the warheads. These improvements have cascaded through to the UK's Trident warhead and RV. The UK government has acknowledged procurement of the Mk4A RV.

Preliminary work on a successor warhead is also underway under the Nuclear Warhead Capability Sustainment Programme at AWE Aldermaston. The replacement submarine will be quieter and stealthier. All of these efforts confirm that the UK continues to be actively engaged in qualitative improvements to its nuclear weapons system.

50. On 2 November 2010, the UK and France concluded a bilateral Treaty for Defence and Security Cooperation. Article 1 of the Treaty provides, *inter alia*:

> The Parties, building on the existing strong links between their respective defense and security communities and armed forces, undertake to build a long-term mutually beneficial partnership in defense and security with the aims of:
>
> ...
> 4. ensuring the viability and safety of their national deterrents, consistent with the Treaty on the Non-Proliferation of Nuclear Weapons.

51. On 18 May 2011, when informing Parliament that the Government had approved the 'Initial Gate' for the nuclear weapons system successor programme, the Secretary of State for Defence explained:

> We have now agreed the broad outline design of the submarine, made some of the design choices-including the propulsion system and the common US-UK missile compartment-and the programme of work we need to start building the first submarine after 2016. We have also agreed the amount of material and parts we will need to buy in advance of the main investment decision. .. Between now and main gate we expect to spend about 15% of the total value of the programme.

52. Although the Secretary of State for Defence denied that the

Government was "locked into any particular strategy before main gate in 2016" and stated that he would "assist the Liberal Democrats in making the case for alternatives" he declared:

> I am absolutely clear that a minimum nuclear deterrent based on the Trident missile delivery system and continuous-at-sea deterrence is right for the United Kingdom and that it should be maintained, and that remains Government policy.

53. On the same day, the Prime Minister told Parliament: "the Government's policy is absolutely clear: we are committed to retaining an independent nuclear deterrent based on Trident".

54. On 30 April 2012, at the First Preparatory Committee for the Ninth Review Conference of the NPT, the Head of the UK Delegation stated:

> As long as large arsenals of nuclear weapons remain and the risk of nuclear proliferation continues, the UK's judgment is that only a credible nuclear capability can provide the necessary ultimate guarantee to our national security. The UK Government is therefore committed to maintaining a minimum national nuclear deterrent, and to proceeding with the renewal of Trident and the submarine replacement programme.

55. On 5 March 2013, in a Statement on Nuclear Disarmament, the UK's Permanent Representative to the Conference on Disarmament declared:

> In 2007, the United Kingdom Parliament debated, and approved by a clear majority, the decision to continue with the programme to renew the UK's nuclear deterrent. The Government set out in the 20 I 0 Strategic Defence and Security Review that the UK would maintain a continuous submarine based deterrent and begin the work of replacing its existing submarines which are due to leave service in the 2020s. This remains the UK Government's policy.

56. On 5 June 2013, in response to a question in Parliament, the Prime Minister stated: "I am strongly committed to the renewal of our deterrent on a like-for-like basis. I think that is right for Britain".

57. The Trident Alternatives Review was published on 16 July 2013. It had been tasked to answer three questions:

> a. Are there credible alternatives to a submarine-based deterrent?
>
> b. Are there credible submarine-based alternatives to the current proposal, *e.g.* Astute with cruise missiles?

c. Are there alternative nuclear postures, i.e. non-continuous-at-sea deterrence ("CASD"), which could maintain credibility?

58. The Trident Alternatives Review concluded: "None of these alternative systems and postures offers the same degree of resilience as the current posture of Continuous at Sea Deterrence, nor could they guarantee a prompt response in all circumstances".

7.3 Views from the UK Trident Deployment Front-Line

An argument used frequently by the UK government to justify its threatened use of nuclear weapons is the current world security situation. Of course words are cheap at the planning stage and politicians are experts at using cheap words to rationalize their policy decisions.

Collectively, the nuclear weapons states (NWS) are responsible for creating a massive security threat to the world. So much so that in a technical sense, because the threats to use nuclear weapons in a range of situations, which is to terrorize potential adversaries, the NWS (including the UN Security Council P5) are the greatest terrorists on the planet.

Unfortunately, it is essential to be concerned with security because, from time to time, rogue groups and governments emerge which are willing to engage in military action against others. This will remain a problem for as long as systems of government enable hotheads to get into positions of power and there remains a thriving market for weapons. It is not necessary to frame all security problems in military terms, because there are other mechanisms in the security arsenal such as non-violence, boycotts, civil disobedience, disruption of supply chains, denial of finance, ostracism, etc.

There are references that help with the debate about the security situation and military defence that does not involve nuclear weapons. Have a look at Robert Green's book *Security Without Nuclear Deterrence* (Green, 2018). Green had operational experience of nuclear weapons as aircrew in Buccaneer nuclear strike jets and anti-submarine helicopters armed with nuclear depth bombs, and as Staff Officer (Intelligence) to CINCFLEET during the 1982 Falklands war in the rank of Commander in the UK's Royal Navy. Before that he worked in the Ministry of Defence. Subsequently his growing concerns about nuclear deterrence led him to become Chair of the UK affiliate of the World Court Project 1991-2004, which resulted in the 1996 International

Court of Justice (ICJ) Advisory Opinion which confirmed the general illegality of the threat, let alone use, of nuclear weapons. The book has important reflections about rationality in policy and decision-making.

Green's book discusses the legal implications of the UK Government narratives which prefer to talk about 'nuclear deterrence' rather than 'nuclear weapons' - a classic information warfare tactic because of the different dimensions of value judgment associated with the words 'deterrent' and 'weapon' - it would be interesting to know which content analysis[2] dictionary was used by those who came up with preferring narrative based on 'deterrence'). Green's book states:

> "**Nuclear Deterrence Implicitly Condemned.** The ICJ, wishing to avoid a direct conflict with the nuclear weapon states, did not specifically pronounce on the legal status of nuclear deterrence. This could have been linked to pressure from nuclear weapon state representatives at the Oral Proceedings on the case. For example, the French delegate warned 'against any pronouncement which, directly or indirectly, might imply judgment being passed on a defence policy based on deterrence.'[i] The UK echoed the US when it said that 'to call in question now the legal basis of the system of deterrence on which so many states have relied for so long for the protection of their peoples could have a profoundly destabilising effect." (Green, 2018; 187-8 - drawing on French and British government oral evidence before the ICJ)

He then draws on the ICJ's Advisory Opinion paragraphs 47 and 105 to reinforce the implicit illegality of nuclear deterrence.

Another important perspective from 'front-line' operations is in Commander Robert Forsyth RN (Ret'd)'s book *Why Trident?* (Forsyth, 2020). Forsyth has command experience in UK submarines used as part of the UK's nuclear deterrence posturing, going as far back as Polaris.

The book presents his findings to answer the question that is the title of his book. In terms of research methodology, he uses secondary data from a variety of available published sources. Primary data includes auto-ethnography to bring in his own experience, observations, reflection and correspondence with a major element of defence policy formation, the UK Ministry of Defence.

The conclusions of the research are stark:

> "...there is no justifiable answer other than national hubris and egotism: a highly dangerous combination that has fuelled 75 years of an arms race that threatens the existence of the world it allegedly protects" (ibid, p 75), and, "the theory of nuclear deterrence is flawed, unproven and poses significant dangers from accidental use" (ibid, back cover).

---

2 For a more detailed understanding of content analysis in this context see Darnton (2005).

The Introduction to Forsyth's book is contributed by UK lawyer Nick Grief who was deeply involved in the RMI v UK case in the ICJ, extensively used in the previous section, and before that in the World Court Project 1991-1996.

For further discussion of Trident and the relevance of Scottish law, see also Zelter (2001), and, Johnson and Zelter (2011).

7.4 The UK 2021 Decision to Increase the Cap on Nuclear Warhead Numbers

In March 2021, the UK Prime Minister sent a shock wave round the world by presenting to Parliament an Integrated Review of Security, Defence, Development and Foreign Policy that announced a substantial increase in the UK stockpile of nuclear weapons. The Integrated Review contained this:

> "In 2010 the Government stated an intent to reduce our overall nuclear warhead stockpile ceiling from not more than 225 to not more than 180 by the mid-2020s. However, in recognition of the evolving security environment, including the developing range of technological and doctrinal threats, this is no longer possible, and the UK will move to an overall nuclear weapon stockpile of no more than 260 warheads." (UK Govt, 2021: p76).

As can be imagined, this caused a flurry of comment and activity by many concerned about this announcement without any accompanying Parliamentary debate, and with such an important statement about increasing the cap on the number of nuclear warheads 'buried' on p76 of the document.

This situation is democratically outrageous and seems to be a classic example of what Lord Hailsham described in his Richard Dimbleby lecture on BBC 1 on 14th October 1976, when describing the British Government as perilously close to an "elective dictatorship" with the executive controlling the legislature, not the other way round! (not available at the BBC at the time of writing).

The Campaign for Nuclear Disarmament was quick off the mark and commissioned a joint legal opinion on the legality of the Integrated Review with reference to nuclear weapons. That was provided in April 2021 (Chinkin and Arimatsu, 2021). This is an important document that deserves careful reading. Its key conclusions are:

> 138. In our opinion, for the reasons set out above:
> (i) The announcement by the UK government of the increase in nuclear warheads and its modernisation of its weapons system constitutes a breach

of the NPT article VI;

(ii) The UK would be in breach of international law were it to use or threaten to use nuclear weapons against a state party to the NPT solely on the basis of a material breach of the latter's non-proliferation obligations;

(iii) The UK would be in breach of international law were it to use or threaten to use nuclear weapons in self-defence solely on the grounds that the future threat of weapons of mass destruction, such as chemical and biological capabilities or emerging technologies, could have comparable impact to nuclear weapons.

The Joint Opinion makes careful relevant use of many points in the ICJ 1996 AO as required by the AO para 104.

By putting the UK in material breach of the NPT, the UK could be putting the whole NPT in jeopardy by virtue of the Vienna Convention on the Law of Treaties. This possibility renders the UK Government action to be reckless.

There are other key points additional to the Joint Opinion: (1) the Integrated Review is also in breach of the Nuremberg Principles, which means that all the individuals involved in developing, approving, and implementing the Review, certainly in terms of nuclear weapons and related policy, attract individual and personal liability under international criminal law; (2) the UK is not 'merely' a State Party to the NPT; it is also a Depositary Government for the NPT which imposes further state responsibilities including cases where differences arise between a Depositary Government and fundamental objectives of the treaty (VCLT, 1969; Rosenne, 1967).

Furthermore, the UK Government's 2021 Integrated Review poses a substantial change in policy, reversing many assurances given publicly and to a variety of international bodies over several decades. This has been done without any consultation, public, or Parliamentary discussion.

For examples of UK groups focussed on Trident, see PICAT who would like to see Tridents before the courts (their website has many useful documents), and the Campaign for Nuclear Disarmament.

7.5 Closing Comments

The UK pursuit of activities that are gravely unlawful under international law goes right back to the birth of the atomic bomb and the bombings of Hiroshima and Nagasaki. All stages have involved working closely with the USA's nuclear weapons capability. The UK stance of nuclear 'deterrence' involves a continuous illegal threat to use nuclear weapons. It has continued a decades long cynical disregard for its legal obligations under the NPT whilst hypocritically paying lip service to its importance. The 2021 Integrated Review is a material breach of the NPT which could pose an existential threat to the NPT itself, thus worsening the very security situation of the UK it claims to be enhancing.

Chapter 8

Civil Society Manifesto to Move Forward

8.1 Introduction

This chapter brings together a summary of ideas that individuals, civil society, diplomats, and countries can pick up to move forward from the TPNW. Different people like to work on different things, so choose what suits you best - and you may have other ideas also. Many things are needed for Moving Forward from the TPNW.

8.2 TPNW

122 countries voted in favour of the TPNW in UNGA in 2017. Following that strong vote in favour of TPNW, 86 countries have signed it after it was open for signature. It entered into for on 22nd January 2021, 90 days after the 50th ratification on 24th October 2020. At the time of this book's publication, 54 countries have ratified TPNW. You can track the progress of ratifications at: https://www.icanw.org/signature_and_ratification_status.

Many countries came under enormous diplomatic pressure from nuclear weapons states not to sign or ratify TPNW.

Clearly, there are more steps to be taken:
1. Encourage states who signed TPNW to ratify it if they have not already done so.
2. Encourage those states who voted in favour of TPNW to join it
3. Encourage all states who have not yet signed or ratified TPNW to join TPNW and become a State Party.

Every new State Party to the TPNW strengthens and reinforces the NPT, and can create a new nuclear weapons free zone.

8.3 Nuremberg Principles and Deterrence of Individual

Decisions to acquire, finance, research, develop, design, build, deploy, maintain, threaten to use, use nuclear weapons are taken by individuals. Who are they? — find, identify, and document them.

Nuclear weapons are very expensive and making them is very profitable — who are the individuals, banks, institutions, pension funds, state bonds, state investment funds in providing the finance? Find, identify, and document them.

The Nuremberg Principles and Rome Statute provide frameworks for individual responsibilities under international criminal law. See Stewart (2014) for an example of applying the Nuremberg Principles to the Vietnam War.

TPNW helps a lot to suppress these activities by individuals Campaign tirelessly to remove from office, public or private, anyone who has a state of mind willing to do something that could contribute to mass murder — they are mentally not fit to have any management or public responsibility.

Countries that have not joined TPNW can introduce domestic legislation to prohibit involvement in any activities that contribute to nuclear weapons. Military manuals can prohibit all nuclear weapons activities and make clear that nuclear weapons are unlawful and that 'obeying orders' is not a valid defence under the Nuremberg Principles or Rome Statute.

...and see Section 6.4...

8.4 Strengthen the Rome Statute and ICC

The key problem about individual liability under international law for involvement with nuclear weapons is enforcement. This is also true for many other aspects of international law.

The easiest way to solve this problem is the introduction of domestic law to bring international law into domestic legislation. This is a requirement of some treaties, but not all. Of course , power hungry individuals, and political parties that wish to 'hijack' a country by devices such as single party states, first past the post voting systems, or coups d'etat, will resist very strongly attempts to bring international law into domestic law. Dictators and tyrannical political parties prefer prerogative powers over domestic legislation.

TPNW requires domestic implementation of treaty prohibitions. The nuremberg Principles and crimes according to the Rome Statute need to be brought into all domestic legislation. This gives an important target for national civil society organizations.

8.5 Nuclear Weapons Convention

Implementing the multilateral nuclear disarmament called for by the Non-Proliferation Treaty will require a new international instrument.

Work on such an instrument has been going on for many years with ideas and drafts for a Nuclear Weapons Convention.. The last draft considered by UNGA was in 2007. That is included in this book as Appendix B. As a result of the passage of time, it will need at lest updating, but it provides a solid basis to commence negotiations. For earlier similar drafts, see Datan and Ware (1999) and, Datan, Hill, Scheffran and Ware (2007).

A coordinated international civil society push for this could help to break the 50+ year NPT deadlock.

8.6 Tribunal on Lawfulness of Hiroshima and Nagasaki

The Allies were the only organizers of the Nuremberg and Tokyo Tribunals, after WWII.

The Allies only put on trial prominent people from the Axis power - what can be described as 'Victors' Justice'.s

Some actions of the Allies, such as carpet or incendiary bombing of large civilian populations and the atomic bombings of Hiroshima and Nagasaki were, *prima facie*, grave breaches of international law as it existed at the time. Individuals of the Allied powers have never been called to account for such decisions and actions.

Even though likely to be controversial, it would be helpful for lawyers and civil society to organize Tribunals, similar to the London Nuclear Warfare and Australia Tribunals, to examine the conduct of the Allies during WWII.

8.7 Change the Narrative Around the NPT

So much of the narrative around the NPT is driven by the NWS states in terms of 'internation security', 'keeping the peace for more than 50 years', and so forth.

TPNW was born in part from the frustration of so many individuals, civil society, and non nuclear weapon states at the NWS impasse on their legal obligations under the NPT.

The recent announcements by NWS of modernizes to nuclear weapons systems combined with states such as Iran and DPRK are posing serious threats to the continuation of the NPT.

The narrative need to move to re-emphasizing the failure of the NWS (abd other states) to insist on the implementation of Article VI..

Ask all parties, frequently, when will, they act in accordance with the NPT?

8.8 Change the Narrative Around Deterrence

Threatening to use weapons where the use of thw eapons would be unlawful, is itself unlawful. Nuclear deterrence is unlawful.

Every time anyone uses the term 'nuclear deterrence' point out that is an admission of illegality.

8.9 Change the Narrative Around Nuclear Reactors

The original and primary purpose of nuclear reactors was to create the fuels needed by nuclear weapons. Using waste heat to generate electricity was a by-product.

Reactors have escaped prohibition under treaties by classifying them as 'peaceful' uses of nuclear energy. Call them out for what they are: generators of nuclear weapons fuels.

There is the additional serious long-term prob;em that the safe disposal of the mounting stockpiles of waste nuclear materials has never been solved posing huge costs and dangers to future generations.

Call for closing down all nuclear reactors and include them in treaties.

8.10 Establish a TPNW for Individuals - TPNWI

Treaties are viewed currently as devices for states. There is no reason why civil society could not prepare a variant of the TPNW for signature and ratification by individuals who would agree to a set of prohibitions that apply to their own conduct in life. It could even have provision for states to accede to it.

Such a Treaty could be entitled something like **Treaty on the Prohibition of Nuclear Weapons for or between Individuals.**

8.11 Further Ideas for Moving Forward

Many other ideas for Moving Forward have been mentioned, including: be vigilant and do something - each person can pick something to do to take TPNW forward; ratifying and respecting other treaties and agreements; alternatives to deterrence and boost the Green Economy; World Order, nation states, sovereignty and peace - bring an end to the ability of sovereign states to hide behind their sovereignty to place the planet and humanity in peril; create a Nuclear Weapons Register of individuals and organizations involved with nuclear weapons.

Treaty on the Prohibition of Nuclear Weapons, 2017

Preamble

The States Parties to this Treaty,

Determined to contribute to the realization of the purposes and principles of the Charter of the United Nations,

Deeply concerned about the catastrophic humanitarian consequences that would result from any use of nuclear weapons, and recognizing the consequent need to completely eliminate such weapons, which remains the only way to guarantee that nuclear weapons are never used again under any circumstances,

Mindful of the risks posed by the continued existence of nuclear weapons, including from any nuclear-weapon detonation by accident, miscalculation or design, and emphasizing that these risks concern the security of all humanity, and that all States share the responsibility to prevent any use of nuclear weapons,

Cognizant that the catastrophic consequences of nuclear weapons cannot be adequately addressed, transcend national borders, pose grave implications for human survival, the environment, socioeconomic development, the global economy, food security and the health of current and future generations, and have a disproportionate impact on women and girls, including as a result of ionizing radiation,

Acknowledging the ethical imperatives for nuclear disarmament and the urgency of achieving and maintaining a nuclear-weapon-free world, which is a global public good of the highest order, serving both national and collective security interests,

Mindful of the unacceptable suffering of and harm caused to the victims of the use of nuclear weapons (hibakusha), as well as of those affected by the testing of nuclear weapons,

Recognizing the disproportionate impact of nuclear-weapon activities on indigenous peoples,

Reaffirming the need for all States at all times to comply with applicable international law, including international humanitarian law and international human rights law,

Basing themselves on the principles and rules of international humanitarian law, in particular the principle that the right of parties to an armed conflict to choose methods or means of warfare is not unlimited, the rule of distinction, the prohibition against indiscriminate attacks, the rules on proportionality and precautions in attack, the prohibition on the use of weapons of a nature to cause superfluous injury or unnecessary suffering, and the rules for the protection of the natural environment,

Considering that any use of nuclear weapons would be contrary to the rules of international law applicable in armed conflict, in particular the principles and rules of international humanitarian law,

Reaffirming that any use of nuclear weapons would also be abhorrent to the principles of humanity and the dictates of public conscience,

Recalling that, in accordance with the Charter of the United Nations, States must refrain in their international relations from the threat or use of force against the territorial integrity or political independence of any State, or in any other manner inconsistent with the Purposes of the United Nations, and that the establishment and maintenance of international peace and security are to be promoted with the least diversion for armaments of the world's human and economic resources,

Recalling also the first resolution of the General Assembly of the United Nations, adopted on 24 January 1946, and subsequent resolutions which call for the elimination of nuclear weapons,

Concerned by the slow pace of nuclear disarmament, the continued reliance on nuclear weapons in military and security concepts, doctrines and policies, and the waste of economic and human resources on programmes for the production, maintenance and modernization of nuclear weapons,

Recognizing that a legally binding prohibition of nuclear weapons constitutes an important contribution towards the achievement and maintenance of a world free of nuclear weapons, including the irreversible, verifiable and transparent elimination of nuclear weapons, and determined to act towards that end,

Determined to act with a view to achieving effective progress towards general and complete disarmament under strict and effective international control,

Reaffirming that there exists an obligation to pursue in good faith and bring to a conclusion negotiations leading to nuclear disarmament in all its aspects under strict and effective international control,

Reaffirming also that the full and effective implementation of the Treaty on the Non-Proliferation of Nuclear Weapons, which serves as the cornerstone of the nuclear disarmament and non-proliferation regime, has a vital role to play in promoting international peace and security,

Recognizing the vital importance of the Comprehensive Nuclear-Test-Ban Treaty and its verification regime as a core element of the nuclear disarmament and non-proliferation regime,

Reaffirming the conviction that the establishment of the internationally recognized nuclear-weapon-free zones on the basis of arrangements freely arrived at among the States of the region concerned enhances global and regional peace and security, strengthens the nuclear non-proliferation regime and contributes towards realizing the objective of nuclear disarmament,

Emphasizing that nothing in this Treaty shall be interpreted as affecting the inalienable right of its States Parties to develop research, production and use of nuclear energy for peaceful purposes without discrimination,

Recognizing that the equal, full and effective participation of both women and men is an essential factor for the promotion and attainment of sustainable peace and security, and committed to supporting and strengthening the effective participation of women in nuclear disarmament,

Recognizing also the importance of peace and disarmament education in all its aspects and of raising awareness of the risks and consequences of nuclear weapons for current and future generations, and committed to the dissemination of the principles and norms of this Treaty,

Stressing the role of public conscience in the furthering of the principles of humanity as evidenced by the call for the total elimination of nuclear weapons, and recognizing the efforts to that end undertaken by the United Nations, the International Red Cross and Red Crescent Movement, other international and regional organizations, non-governmental organizations, religious leaders, parliamentarians, academics and the hibakusha,

Have agreed as follows:

Article 1 - Prohibitions

1. Each State Party undertakes never under any circumstances to:

(a) Develop, test, produce, manufacture, otherwise acquire, possess or stockpile nuclear weapons or other nuclear explosive devices;

(b) Transfer to any recipient whatsoever nuclear weapons or other nuclear explosive devices or control over such weapons or explosive devices directly or indirectly;

(c) Receive the transfer of or control over nuclear weapons or other nuclear explosive devices directly or indirectly;

(d) Use or threaten to use nuclear weapons or other nuclear explosive devices;

(e) Assist, encourage or induce, in any way, anyone to engage in any activity prohibited to a State Party under this Treaty;

(f) Seek or receive any assistance, in any way, from anyone to engage in any activity prohibited to a State Party under this Treaty;

(g) Allow any stationing, installation or deployment of any nuclear weapons or other nuclear explosive devices in its territory or at any place under its jurisdiction or control.

Article 2 - Declarations

1. Each State Party shall submit to the Secretary-General of the United Nations, not later than 30 days after this Treaty enters into force for that State Party, a declaration in which it shall:

(a) Declare whether it owned, possessed or controlled nuclear weapons or nuclear explosive devices and eliminated its nuclear-weapon programme, including the elimination or irreversible conversion of all nuclear-weapons-related facilities, prior to the entry into force of this Treaty for that State Party;

(b) Notwithstanding Article 1 (a), declare whether it owns, possesses or controls any nuclear weapons or other nuclear explosive devices;

(c) Notwithstanding Article 1 (g), declare whether there are any nuclear weapons or other nuclear explosive devices in its territory or in any place under its jurisdiction or control that are owned, possessed or controlled by another State.

2. The Secretary-General of the United Nations shall transmit all such declarations received to the States Parties.

Article 3 - Safeguards

1. Each State Party to which Article 4, paragraph 1 or 2, does not apply shall, at a minimum, maintain its International Atomic Energy Agency safeguards obligations in force at the time of entry into

force of this Treaty, without prejudice to any additional relevant instruments that it may adopt in the future.

2. Each State Party to which Article 4, paragraph 1 or 2, does not apply that has not yet done so shall conclude with the International Atomic Energy Agency and bring into force a comprehensive safeguards agreement (INFCIRC/153 (Corrected)). Negotiation of such agreement shall commence within 180 days from the entry into force of this Treaty for that State Party. The agreement shall enter into force no later than 18 months from the entry into force of this Treaty for that State Party. Each State Party shall thereafter maintain such obligations, without prejudice to any additional relevant instruments that it may adopt in the future.

Article 4 - Towards the total elimination of nuclear weapons

1. Each State Party that after 7 July 2017 owned, possessed or controlled nuclear weapons or other nuclear explosive devices and eliminated its nuclear-weapon programme, including the elimination or irreversible conversion of all nuclear-weapons-related facilities, prior to the entry into force of this Treaty for it, shall cooperate with the competent international authority designated pursuant to paragraph 6 of this Article for the purpose of verifying the irreversible elimination of its nuclear-weapon programme. The competent international authority shall report to the States Parties. Such a State Party shall conclude a safeguards agreement with the International Atomic Energy Agency sufficient to provide credible assurance of the non-diversion of declared nuclear material from peaceful nuclear activities and of the absence of undeclared nuclear material or activities in that State Party as a whole. Negotiation of such agreement shall commence within 180 days from the entry into force of this Treaty for that State Party. The agreement shall enter into force no later than 18 months from the entry into force of this Treay for that State Party. That State Party shall thereafter, at a minimum, maintain these safeguards obligations, without prejudice to any additional relevant instruments that it may adopt in the future.

2. Notwithstanding Article 1 (a), each State Party that owns, possesses or controls nuclear weapons or other nuclear explosive devices shall immediately remove them from operational status, and destroy them as soon as possible but not later than a deadline to be determined by the first meeting of States Parties, in accordance

with a legally binding, time-bound plan for the verified and irreversible elimination of that State Party's nuclear-weapon programme, including the elimination or irreversible conversion of all nuclear-weapons-related facilities. The State Party, no later than 60 days after the entry into force of this Treaty for that State Party, shall submit this plan to the States Parties or to a competent international authority designated by the States Parties. The plan shall then be negotiated with the competent international authority, which shall submit it to the subsequent meeting of States Parties or review conference, whichever comes first, for approval in accordance with its rules of procedure.

3. A State Party to which paragraph 2 above applies shall conclude a safeguards agreement with the International Atomic Energy Agency sufficient to provide credible assurance of the non-diversion of declared nuclear material from peaceful nuclear activities and of the absence of undeclared nuclear material or activities in the State as a whole. Negotiation of such agreement shall commence no later than the date upon which implementation of the plan referred to in paragraph 2 is completed. The agreement shall enter into force no later than 18 months after the date of initiation of negotiations. That State Party shall thereafter, at a minimum, maintain these safeguards obligations, without prejudice to any additional relevant instruments that it may adopt in the future. Following the entry into force of the agreement referred to in this paragraph, the State Party shall submit to the Secretary-General of the United Nations a final declaration that it has fulfilled its obligations under this Article.

4. Notwithstanding Article 1 (b) and (g), each State Party that has any nuclear weapons or other nuclear explosive devices in its territory or in any place under its jurisdiction or control that are owned, possessed or controlled by another State shall ensure the prompt removal of such weapons, as soon as possible but not later than a deadline to be determined by the first meeting of States Parties. Upon the removal of such weapons or other explosive devices, that State Party shall submit to the Secretary-General of the United Nations a declaration that it has fulfilled its obligations under this Article.

5. Each State Party to which this Article applies shall submit a report to each meeting of States Parties and each review conference on the progress made towards the implementation of its obligations under this Article, until such time as they are fulfilled.

6. The States Parties shall designate a competent international

authority or authorities to negotiate and verify the irreversible elimination of nuclear-weapons programmes, including the elimination or irreversible conversion of all nuclear-weapons-related facilities in accordance with paragraphs 1, 2 and 3 of this Article. In the event that such a designation has not been made prior to the entry into force of this Treaty for a State Party to which paragraph 1 or 2 of this Article applies, the Secretary-General of the United Nations shall convene an extraordinary meeting of States Parties to take any decisions that may be required.

Article 5 - National implementation

1. Each State Party shall adopt the necessary measures to implement its obligations under this Treaty.
2. Each State Party shall take all appropriate legal, administrative and other measures, including the imposition of penal sanctions, to prevent and suppress any activity prohibited to a State Party under this Treaty undertaken by persons or on territory under its jurisdiction or control.

Article 6 - Victim assistance and environmental remediation

1. Each State Party shall, with respect to individuals under its jurisdiction who are affected by the use or testing of nuclear weapons, in accordance with applicable international humanitarian and human rights law, adequately provide age- and gender-sensitive assistance, without discrimination, including medical care, rehabilitation and psychological support, as well as provide for their social and economic inclusion.
2. Each State Party, with respect to areas under its jurisdiction or control contaminated as a result of activities related to the testing or use of nuclear weapons or other nuclear explosive devices, shall take necessary and appropriate measures towards the environmental remediation of areas so contaminated.
3. The obligations under paragraphs 1 and 2 above shall be without prejudice to the duties and obligations of any other States under international law or bilateral agreements.

Article 7 - International cooperation and assistance

1. Each State Party shall cooperate with other States Parties to

facilitate the implementation of this Treaty.

2. In fulfilling its obligations under this Treaty, each State Party shall have the right to seek and receive assistance, where feasible, from other States Parties.

3. Each State Party in a position to do so shall provide technical, material and financial assistance to States Parties affected by nuclear-weapons use or testing, to further the implementation of this Treaty.

4. Each State Party in a position to do so shall provide assistance for the victims of the use or testing of nuclear weapons or other nuclear explosive devices.

5. Assistance under this Article may be provided, inter alia, through the United Nations system, international, regional or national organizations or institutions, non-governmental organizations or institutions, the International Committee of the Red Cross, the International Federation of Red Cross and Red Crescent Societies, or national Red Cross and Red Crescent Societies, or on a bilateral basis.

6. Without prejudice to any other duty or obligation that it may have under international law, a State Party that has used or tested nuclear weapons or any other nuclear explosive devices shall have a responsibility to provide adequate assistance to affected States Parties, for the purpose of victim assistance and environmental remediation.

Article 8 - Meeting of States Parties

1. The States Parties shall meet regularly in order to consider and, where necessary, take decisions in respect of any matter with regard to the application or implementation of this Treaty, in accordance with its relevant provisions, and on further measures for nuclear disarmament, including:

(a) The implementation and status of this Treaty;

(b) Measures for the verified, time-bound and irreversible elimination of nuclear-weapon programmes, including additional protocols to this Treaty;

(c) Any other matters pursuant to and consistent with the provisions of this Treaty.

2. The first meeting of States Parties shall be convened by the Secretary-General of the United Nations within one year of the entry into force of this Treaty. Further meetings of States Parties shall be convened by the Secretary-General of the United Nations on a biennial basis,

unless otherwise agreed by the States Parties. The meeting of States Parties shall adopt its rules of procedure at its first session. Pending their adoption, the rules of procedure of the United Nations conference to negotiate a legally binding instrument to prohibit nuclear weapons, leading towards their total elimination, shall apply.

3. Extraordinary meetings of States Parties shall be convened, as may be deemed necessary, by the Secretary-General of the United Nations, at the written request of any State Party provided that this request is supported by at least one third of the States Parties.

4. After a period of five years following the entry into force of this Treaty, the Secretary-General of the United Nations shall convene a conference to review the operation of the Treaty and the progress in achieving the purposes of the Treaty. The Secretary-General of the United Nations shall convene further review conferences at intervals of six years with the same objective, unless otherwise agreed by the States Parties.

5. States not party to this Treaty, as well as the relevant entities of the United Nations system, other relevant international organizations or institutions, regional organizations, the International Committee of the Red Cross, the International Federation of Red Cross and Red Crescent Societies and relevant non-governmental organizations, shall be invited to attend the meetings of States Parties and the review conferences as observers.

Article 9 - Costs

1. The costs of the meetings of States Parties, the review conferences and the extraordinary meetings of States Parties shall be borne by the States Parties and States not party to this Treaty participating therein as observers, in accordance with the United Nations scale of assessment adjusted appropriately.

2. The costs incurred by the Secretary-General of the United Nations in the circulation of declarations under Article 2, reports under Article 4 and proposed amendments under Article 10 of this Treaty shall be borne by the States Parties in accordance with the United Nations scale of assessment adjusted appropriately.

3. The cost related to the implementation of verification measures required under Article 4 as well as the costs related to the destruction of nuclear weapons or other nuclear explosive devices, and the elimination of nuclear-weapon programmes, including the elimination

or conversion of all nuclear-weapons-related facilities, should be borne by the States Parties to which they apply.

Article 10 - Amendments

1. At any time after the entry into force of this Treaty, any State Party may propose amendments to the Treaty. The text of a proposed amendment shall be communicated to the Secretary-General of the United Nations, who shall circulate it to all States Parties and shall seek their views on whether to consider the proposal. If a majority of the States Parties notify the Secretary-General of the United Nations no later than 90 days after its circulation that they support further consideration of the proposal, the proposal shall be considered at the next meeting of States Parties or review conference, whichever comes first.
2. A meeting of States Parties or a review conference may agree upon amendments which shall be adopted by a positivevote of a majority of two thirds of the States Parties. The Depositary shall communicate any adopted amendment to all States Parties.
3. The amendment shall enter into force for each State Party that deposits its instrument of ratification or acceptance of the amendment 90 days following the deposit of such instruments of ratification or acceptance by a majority of the States Parties at the time of adoption. Thereafter, it shall enter into force for any other State Party 90 days following the deposit of its instrument of ratification or acceptance of the amendment.

Article 11 - Settlement of disputes

1. When a dispute arises between two or more States Parties relating to the interpretation or application of this Treaty, the parties concerned shall consult together with a view to the settlement of the dispute by negotiation or by other peaceful means of the parties' choice in accordance with Article 33 of the Charter of the United Nations.
2. The meeting of States Parties may contribute to the settlement of the dispute, including by offering its good offices, calling upon the States Parties concerned to start the settlement procedure of their choice and recommending a time limit for any agreed procedure, in accordance with the relevant provisions of this Treaty and the Charter of the United Nations.

Article 12 - Universality

Each State Party shall encourage States not party to this Treaty to sign, ratify, accept, approve or accede to the Treaty, with the goal of universal adherence of all States to the Treaty.

Article 13 - Signature

This Treaty shall be open for signature to all States at United Nations Headquarters in New York as from 20 September 2017.

Article 14 - Ratification, acceptance, approval or accession

This Treaty shall be subject to ratification, acceptance or approval by signatory States. The Treaty shall be open for accession.

Article 15 - Entry into force

1. This Treaty shall enter into force 90 days after the fiftieth instrument of ratification, acceptance, approval or accession has been deposited.
2. For any State that deposits its instrument of ratification, acceptance, approval or accession after the date of the deposit of the fiftieth instrument of ratification, acceptance, approval or accession, this Treaty shall enter into force 90 days after the date on which that State has deposited its instrument of ratification, acceptance, approval or accession.

Article 16 - Reservations

The Articles of this Treaty shall not be subject to reservations.

Article 17 - Duration and withdrawal

1. This Treaty shall be of unlimited duration.
2. Each State Party shall, in exercising its national sovereignty, have the right to withdraw from this Treaty if it decides that extraordinary events related to the subject matter of the Treaty have jeopardized the supreme interests of its country. It shall give notice of such withdrawal to the Depositary. Such notice shall include a statement

of the extraordinary events that it regards as having jeopardized its supreme interests.

3. Such withdrawal shall only take effect 12 months after the date of the receipt of the notification of withdrawal by the Depositary. If, however, on the expiry of that 12-month period, the withdrawing State Party is a party to an armed conflict, the State Party shall continue to be bound by the obligations of this Treaty and of any additional protocols until it is no longer party to an armed conflict.

Article 18 - Relationship with other agreements

The implementation of this Treaty shall not prejudice obligations undertaken by States Parties with regard to existing international agreements, to which they are party, where those obligations are consistent with the Treaty.

Article 19 - Depositary

The Secretary-General of the United Nations is hereby designated as the Depositary of this Treaty.

Article 20 - Authentic texts

The Arabic, Chinese, English, French, Russian and Spanish texts of this Treaty shall be equally authentic.

DONE at New York, this seventh day of July, two thousand and seventeen.

Model Nuclear Weapons, Convention

Model Nuclear Weapons Convention

Convention on the Prohibition of the Development, Testing, Production, Stockpiling, Transfer, Use and Threat of Use of Nuclear Weapons and on Their Elimination

April 2007

Updated from the Model Nuclear Weapons Convention circulated in November 1997 as United Nations document A/C.1/52/7

MODEL NUCLEAR WEAPONS CONVENTION

Notes:

1. The Model Nuclear Weapons Convention has been prepared by a consortium of scientists, lawyers, disarmament experts, academics and officials as a discussion document to assist in deliberations and possible negotiations leading to the prohibition and elimination of nuclear weapons. It outlines legal, technical and political elements which could be utilized in an actual nuclear weapons convention or package/framework of agreements. The drafters do not assume that the final agreed convention or package of agreements would be exactly the same as in this Model. The drafters do however believe that this Model demonstrates the feasibility and practicality of nuclear disarmament. For discussion on these issues see *Securing our Survival: The Case for a Nuclear Weapons Convention*, IPPNW, Cambridge USA, 2007.

2. [Square brackets] refers to text which has not been agreed by all the drafters or which suggests alternative approaches.

The text makes reference to a "Verification Annex" which would form an integral part of a negotiated NWC, but is not included in this Model NWC.

Summary of the Model Nuclear Weapons Convention

General Obligations

The Model Nuclear Weapons Convention prohibits development, testing, production, stockpiling, transfer, use and threat of use of nuclear weapons. States possessing nuclear weapons will be required to destroy their arsenals according to a series of phases. The Convention also prohibits the production of weapons usable fissile material and requires delivery vehicles to be destroyed or converted to make them non-nuclear capable.

Declarations

States parties to the Convention will be required to declare all nuclear weapons, nuclear material, nuclear facilities and nuclear weapons delivery vehicles they possess or control, and the locations of these.

Phases for elimination

The Convention outlines a series of five phases for the elimination of nuclear weapons beginning with taking nuclear weapons off alert, removing weapons from deployment, removing nuclear warheads from their delivery vehicles, disabling the warheads, removing and disfiguring the "pits" and placing the fissile material under international control. In the initial phases the U.S. and Russia are required to make the deepest cuts in their nuclear arsenals.

Verification

Verification will include declarations and reports from States, routine inspections, challenge inspections, on-site sensors, satellite photography, radionuclide sampling and other remote sensors, information sharing with other organizations, and citizen reporting. Persons reporting suspected violations of the convention will be provided protection through the Convention including the right of asylum.

An International Monitoring System will be established under the Convention to gather information, and will make most of this information available through a registry. Information which may jeopardize commercial secrets or national security will be kept confidential.

National Implementation Measures

States parties are required to adopt necessary legislative measures to implement their obligations under the Convention to provide for prosecution of persons committing crimes and protection for persons reporting violations of the Convention.

States are also required to establish a national authority to be responsible for national tasks in implementation.

Rights and Obligations of Persons

The Convention applies rights and obligations to individuals and legal entities as well as States. Individuals have an obligation to report violations of the Convention and the right to protection if they do so.

Procedures for the apprehension and fair trial of individuals accused of committing crimes under the treaty are provided.

Agency

An agency would be established to implement the Convention. It will be responsible for verification, ensuring compliance, and decision making, and will comprise a Conference of States Parties, an Executive Council and a Technical Secretariat.

Nuclear Material

The Convention prohibits the production of any fissionable or fusionable material which can be used directly to make a nuclear weapon, including plutonium (other than that in spent fuel) and highly enriched uranium. Low enriched uranium would be permitted for nuclear energy purposes.

Cooperation, Compliance and Dispute Settlement

Provisions are included for consultation, cooperation and fact-finding to clarify and resolve questions of interpretation with respect to compliance and other matters. A legal dispute may be referred to the International Court of Justice by mutual consent of States Parties. The Agency may also recommend to the United Nations General Assembly to request an advisory opinion from the ICJ on a legal dispute.

The Convention provides for a series of graduated responses for non-compliance beginning with consultation and clarification, negotiation, and, if required, sanctions or recourse to the U.N. General Assembly and Security Council for action.

Relation with other international agreements

The Model NWC would build on existing nuclear nonproliferation and disarmament regimes and verification and compliance arrangements, including the Non-Proliferation Treaty, International Atomic Energy Agency Safeguards, Comprehensive Test Ban Treaty Organisation International Monitoring System and bilateral agreements between Russia and the United States. In some cases the NWC may add to the functions and activities of such regimes and arrangements. In other cases, theNWC would establish additional complementary arrangements.

Financing

Nuclear weapon states are obliged to cover the costs of the elimination of their nuclear arsenals. However, an international fund will be established to assist states that may have financial difficulties in meeting their obligations.

Optional Protocol Concerning Energy Assistance

The Convention does not prohibit the use of nuclear energy for peaceful purposes. However, it includes an optional protocol which would establish a program of energy assistance for States parties choosing not to develop nuclear energy or to phase out existing nuclear energy programs.

Preamble

We the people of the Earth, through the States Parties to this Convention:

Convinced that the existence of nuclear weapons poses a threat to all humanity and that their use would have catastrophic consequences for all the creatures of this Earth;

Noting that the destructive effects of nuclear weapons upon life on earth are uncontrollable whether in time or space;

Aware that amongst weapons of mass destruction, the abolition of which is recognized as being in the collective security interest of all people and States, nuclear weapons are unprecedented an unequalled in destructive potential;

Affirming that the inherent dignity and equal and inalienable rights of all members of the huma family include the right to life, liberty, peace and the security of person;

Convinced that all countries have an obligation to make every effort to achieve the goal of eliminatin nuclear weapons, the terror which they hold for humankind and the threat which they pose to life on Earth;

Recognizing that numerous regions, including Antarctica, Outer Space, Latin America, the Sea Bed, the South Pacific, Southeast Asia,

Africa, and Central Asia have already been established as nuclear weapon free zones, where possession, production, development, deployment, use and threat of use of nuclear weapons are forever prohibited, and desiring to extend this benefit to the entire planet for the good of all life;

Determined to eliminate the risks of environmental pollution by radioactive waste and other radioactive matter associated with nuclear weapons and to ensure that the bounty and beauty of the Earth shall remain the common heritage of all of us and our descendants in perpetuity to be enjoyed by all in peace;

Recognizing the universal need for environmentally safe, sustainable energy;

Gravely concerned that the use of nuclear weapons may be brought about not only intentionally by war or terrorism, but also through human or mechanical error or failure, and that the very existence and gravity of these threats of nuclear weapons use generates a climate of suspicion and fear which is antagonistic to the promotion of universal respect for and observance of the human rights and fundamental freedoms set forth in the Charter of the United Nations and the Universal Declaration of Human Rights;

Convinced of the serious threats posed to the environment by nuclear arsenals, the economic and social costs and waste of intellectual talent occasioned by these arsenals and the efforts required to prevent their use, the dangers inherent in the existence of the materials used to make nuclear weapons and the attendant problems of proliferation, the medically and psychologically catastrophic effects of any use of a nuclear weapon, the potential effects of mutations on the genetic pool and numerous other risks associated with nuclear weapons;

Welcoming the Convention on the Prohibition of the Development, Production and Stockpiling of Bacteriological (Biological) and Toxin Weapons and on Their Destruction and the Convention on the Prohibition of the Development, Production, Stockpiling and Use of Chemical Weapons and on Their Destruction, as indications of a progression toward the elimination of all weapons of mass destruction;

Recognizing that all life is sacred and that there is a moral imperative to eliminate all weapons of mass destruction;

Welcoming the Convention on the Prohibition of the Use, Stockpiling, Production and Transfer of Anti-Personnel Mines and on Their Destruction, as an indication of progress towards the prohibition and elimination of weapons which are indiscriminate and cause unnecessary suffering;

Welcoming also the Rome Statute of the International Criminal Court, in particular the recognition of individual responsibility for crimes involved in employing weapons which cause unnecessary suffering or which are inherently indiscriminate;

Believing that the threat and use of nuclear weapons is incompatible with civilized norms, standards of morality and humanitarian law which prohibit the use of inhumane weapons and those with indiscriminate effects;

Recalling Resolution 1(I), adopted unanimously on January 24, 1946 at the First Session of the General Assembly of the United Nations, and the many subsequent resolutions of the United Nations which call for the elimination of atomic weapons;

Recalling also the Final Document of the United Nations First Special Session of the General Assembly on Disarmament 1978, which calls for the elimination of nuclear weapons;

Mindful of the solemn obligations of States made in Article VI of the Treaty on the Non-Proliferation of Nuclear Weapons to end the nuclear arms race at an early date and achieve nuclear disarmament,

and to further commitments on specific steps to achieve nuclear disarmament in the "Principles and Objectives for Nuclear Non-Proliferation and Disarmament" " agreed in 1995, and the "Practical steps for the systematic and progressive efforts to implement Article VI of the Treaty on the Non-Proliferation of Nuclear Weapons" agreed in 2000;

Convinced that the elimination of nuclear weapons is an important step towards the goal of general and complete disarmament;

Welcoming the advisory opinion of the International Court of Justice of July 8, 1996, which concluded "that the threat or use of nuclear weapons would generally be contrary to the rules of international law applicable in armed conflict, and in particular the principles and rules of humanitarian law", and concluded unanimously that "There exists an obligation to pursue in good faith and bring to a conclusion negotiations leading to nuclear disarmament in all its aspects under strict and effective international control";

Recalling United Nations General Assembly resolutions 51/45 M, of 10 December 1996, 52/38 O of 9 December 1997, 53/77 W of 4 December 1998, 54/54 Q of 1 December 1999, 55/33 X of 20 November 2000, 56/24 S of 29 November 2001, 57/85 of 22 November 2002, 58/46 of 8 December 2003, 59/83 of 3 December 2004, 60/76 of

8 December 2005, and 61/83 of 6 December 2006 which underline the nuclear disarmament obligation affirmed by the International Court of Justice and call "upon all States to fulfil that obligation immediately by commencing multilateral negotiations ... leading to an early conclusion of a nuclear weapons convention prohibiting the development, production, testing, deployment, stockpiling, transfer, threat or use of nuclear weapons and providing for their elimination";

Convinced that a convention prohibiting the development, testing, production, stockpiling, transfer, use and threat of use of nuclear weapons and providing for their elimination is required to abolish these weapons from the Earth;

Have agreed as follows:

I. General Obligations

A. State Obligations

1. Each State Party to this Convention undertakes never under any circumstances:

a. To use or threaten to use nuclear weapons;

b. To engage in any military or other preparations to use nuclear weapons;

c. To develop, test, produce, otherwise acquire, deploy, stockpile, maintain, retain, or transfer nuclear weapons except as specified under paragraph 4 of this Article;

d. To develop, test, produce, otherwise acquire, stockpile, retain, transfer or use proscribed nuclear material except as specified under paragraph 4 of this Article;

e. To develop, test, produce, otherwise acquire, deploy, stockpile, maintain, retain, or transfer nuclear weapons delivery vehicles;

f. To develop, test, produce, otherwise acquire, stockpile, maintain, retain, or transfer nuclear weapon components or equipment as specified in this Convention;

g. To fund [or conduct] nuclear weapons research, with the exception of nuclear disarmament research;

h. To assist, encourage, induce or permit, in any way, directly or indirectly, anyone to engage in any activity prohibited under this Convention.

2. Each State Party undertakes:

a. To destroy all nuclear weapons it owns or possesses, or that are located in any place under its jurisdiction or control, in accordance with the provisions of this Convention;

    b. To destroy all nuclear weapons it abandoned on the territory of another State, in accordance with the provisions of this Convention;

    c. To submit all nuclear facilities to preventive controls;

    d. To destroy all nuclear weapons facilities it owns or possesses, or that are located in any place under its jurisdiction or control, or to convert such facilities to weapons destruction facilities or other facilities not prohibited by this Convention;

    e. [To disable or destroy all facilities, systems or sub-systems designed or used in the command or control of nuclear weapons, or convert such facilities, systems or subsystems to purposes not prohibited under this Convention;]

    f. To destroy or convert for purposes not prohibited under this Convention all nuclear weapons delivery vehicles and nuclear weapon components;

    g. To place all special nuclear material under preventive controls as specified in this Convention;

    h. To participate in good faith in activities aimed at the promotion of transparency with respect to nuclear weapons and related technologies and the promotion of education for the purposes of detecting and preventing activities prohibited under this Convention;

    i. To report violations of this Convention to the Agency [and to cooperate to the fullest with the Agency's investigative, monitoring and verification functions;] [and to provide to the Agency all information requested by the Agency for the purposes of implementing this Convention, except such information as may be with-held for legitimate international or national security or trade secret concerns;]

    j. To enact all domestic legislation necessary for the implementation of this Convention.

3. These obligations shall apply equally to nuclear explosive devices intended for peaceful purposes.

4. These obligations shall not be interpreted to prohibit activities consistent with the application and implementation of the provisions of this Convention [including but not limited to transfer of nuclear weapons, special nuclear material, and nuclear weapons delivery vehicles for the purpose of their destruction or disposal, and nuclear disarmament research and verification thereof].

B. Obligations of Persons

5. The following acts are crimes for which persons shall be held responsible regardless of their position, residence, citizenship or country of incorporation:

　　a. To engage or attempt to engage in any acts listed in subparagraphs 1.a through 1.g, inclusive, of this Article;

　　b. To aid, abet, or otherwise assist, in any way, anyone to engage in any activity prohibited under this Convention.

6. The fact that the present Convention provides criminal responsibility for individuals does not affect the responsibility of States under international law.

II. Definitions

A. States and Persons

1. "Nuclear Weapons State" means a state which has manufactured and exploded a nuclear weapon or other nuclear explosive device prior to 1 January 1967.

2. "Nuclear Capable State" means a State which has developed or has the capacity to develop nuclear weapons and which is not party to the Non-Proliferation Treaty.

3. "Person" means a natural or legal person.

B. Nuclear Weapons

4. "Nuclear Weapon" means:

　　a. Any device which is capable of releasing nuclear energy in an uncontrolled manner and which has a group of characteristics that are appropriate for use for warlike purposes;

　　b. Any nuclear explosive device;

　　c. Any radiological weapon; or

　　d. Any weapon which is designed to include a nuclear explosive device as a trigger or other component.

5. "Nuclear Weapon Component" means any constituent part of a nuclear weapon [excluding the special nuclear material when separated from other components].

6. "Nuclear Weapons Delivery Vehicle" means any vehicle designed for or capable of delivering a nuclear weapon. Any nuclear weapons

delivery vehicle that has been constructed, developed, flight-tested or deployed for weapon delivery shall be considered a nuclear weapons delivery vehicle.

7. "Plutonium Pit" means the core element of a nuclear weapon's primary or fission component.

8. "Radiological Weapon" means any weapon that disperses radioactive material or uses radioactive material as a primary material in its construction.

9. "Warhead" means the explosive part of a nuclear weapons system. Warheads consist of nuclear materials, conventional high explosives, related firing mechanisms and containment structure.

C. Nuclear Energy, Explosion, and Explosive Device

10. "Nuclear Energy" means energy released from the nucleus of an atom either spontaneously or through interaction with other particles and/or electromagnetic radiation.

11. "Nuclear Explosion" means the release of significant amounts of nuclear energy on a time-scale faster than or comparable to chemical explosives [including micro-fission, micro-fusion or miniaturized devices of any yield].

12. "Nuclear Explosive Device" means any device capable of undergoing a nuclear explosion, irrespective of its purpose. The term includes such a weapon or device in unassembled and partly assembled forms, as well as devices or assemblies which belong to a nuclear explosive device or are a modification of such suitable for development and testing of nuclear weapons or other nuclear explosive devices, but does not include the means of transport or delivery of such a weapon or device if separable from and not an indivisible part of it.

13. "Significant Amount of Nuclear Energy" means more than the energy released by radioactive decay and spontaneous fission and may be much smaller than the maximum energy yield of the largest chemical explosions.

D. Nuclear Material

14. "Nuclear Material" means any source or fissionable or fusionable material as defined in this Convention.

15. "Exemption Quantities" mean quantities of nuclear material not prohibited under the provisions of this Convention [and preventive controls].

16. "Fissionable Material" means any isotope which may undergo either spontaneous fission or fission induced by neutrons of any energy, as well as any compound ormixture including such isotopes.

17. "Fusionable Material" means any isotope capable of undergoing fusion with the same kind of nuclide or with any other nuclide by applying sufficient conditions (pressure, temperature and inclusion time) with technical means.

18. "Highly Enriched Uranium (HEU)" means uranium in which the naturally occurring U-235 isotope (0.7% in natural uranium) is increased to 20% U-235 or above.

19. "Low Enriched Uranium (LEU)" means uranium enriched in the isotopic content of U-235 but to less than 20% of the total mass.

20. "Mixed Oxide Fuel (MOX fuel)" means nuclear reactor fuel composed of plutonium and uranium oxides.

21. ["Other Special Nuclear Material" means special nuclear material other than plutonium and uranium enriched to 20% or more U-235 or U-233.]

22. "Proscribed fissionable material" means any fissionable material that can be used for the manufacture of nuclear weapons without transmutation, chemical reprocessing or further enrichment, and includes any isotopic mixture of separated and unirradiated plutonium, uranium enriched in the isotoptes 235 to 20% or more, uranium-233.

23. "Proscribed fusionable material" means any fusionable material that can be used for the manufacture of nuclear weapons without transmutation, redoxation or further enrichment.

24. "Proscribed nuclear material" means any proscribed fissionable or any proscribed fusionable material.

25. "Significant quantity" means the approximate quantity of nuclear material in respect of which, taking into account any conversion process involved, the possibility of manufacturing a nuclear explosive device cannot be excluded.

26. "Source Material" means uranium containing the mixture of isotopes occurring in nature; uranium depleted in the isotope U-235, thorium, lithium beyond naturally occurring concentration, deuterium, helium-3, or any of the foregoing in the form of metal, alloy, chemical compound or concentrate.

27. "Special Fissionable Material" means fissionable material that can be used for the manufacture of nuclear weapons.

28. "Special Fusionable Material" means any fusionable material that can be used for the manufacture of nuclear weapons and includes

deuterium, tritium, helium-3, and lithium-6.
29. "Special Nuclear Material" means any special fissionable or any special fusionablematerial.

E. Nuclear Facilities

30. "Nuclear Facility" means any facility for the research, testing, production, extraction, enrichment, processing, reprocessing, or storage of nuclear material; any facility for the production of nuclear energy; any facility for the research, development, testing, production, storage, assembly, disassembly, maintenance, modification, deployment, or delivery of nuclear weapons, or nuclear weapon components; or any facility deemed# a nuclear facility by the Technical Secretariat. The term "Nuclear Facility" includes[but is not limited to] the following:
31. "Command, Control or Communication Facility", means [any facility designed or used for the purpose of launching, targeting, directing or detonating a nuclear weapon or its delivery vehicle, or for aiding or assisting in any of these purposes.]
32. "Deployment Site" means the location where a nuclear weapon is or has been deployed, or a location which is equipped for the deployment of nuclear weapons.
33. "Nuclear Enrichment Facility" means a facility capable of increasing the ratio of the isotope uranium-235 in natural uranium.
34. "Nuclear Material Storage Facility" means a facility for the interim or long-term storage of nuclear material.
35. "Nuclear Reactor" means any device in which a controlled, self-sustaining fission chain-reaction can be maintained or in which a controlled fission chain is maintained partly by an external source of neutrons.
36. "Nuclear Reprocessing Facility" means a facility to separate irradiated nuclear material and fission products in whole or in part, and includes the facility's head-end treatment section and its associated storage and analytical sections.
37. "Nuclear Weapons Destruction Facility" means any facility for disassembly or destruction of nuclear weapons or for rendering them permanently inoperable.
38. "Nuclear Weapons Facility" means any facility for the design, research, development, testing, production, storage, assembly, maintenance, modification, deployment, delivery, command, or control of nuclear weapons or Schedule 1 or Schedule 2 nuclear weapon components.

39. "Nuclear Weapons Production Facility" means any nuclear facility which produces materials which have been or may be used for military purposes, including such a reactor, a plant for processing nuclear material irradiated in a reactor, a plant for separating the isotopes of nuclear material, a plant for processing or fabricating nuclear material, a plant for the construction or assembly of nuclear weapon components, or a facility or plant of such other type as may be deemed a Nuclear Weapons Production Facility by the Technical Secretariat.

40. "Nuclear Weapons Research Facility" means any facility in which nuclear weapons research, development, testing or computer simulation is conducted.

41. "Nuclear Weapons Storage Facility" means a facility for the storage of nuclear weapons but does not include such a facility located on a deployment site.

42. "Nuclear Weapons Testing Facility" means a facility or prepared site for conducting nuclear weapons testing.

F. Nuclear Activities

43. "Nuclear Activity" means:

    a. Any construction or use of a nuclear reactor or component parts thereof;

    b. Any production, use or threat of use of a nuclear weapon;

    c. Any research, development or testing of nuclear energy or nuclear weapons;

    d. Any production, separation, treatment or handling of nuclear material;

    e. Any dismantling, disabling or destruction of nuclear weapons;

    f. Any decommissioning of nuclear reactors and power plants;

    g. Any application of radiation and isotopes in food, agriculture, med engineering, geology or other industrial processes; or

    h. Any other activity listed below or deemed a nuclear activity by the Agency.

44. "Convert" means modify to a use not prohibited under this Convention.

45. "De-alert" means reduce the alert status of nuclear weapons by eliminating launch-onwarning or launch-under-attack alert readiness postures, e.g., by removing key trigger mechanisms, decoupling warheads from nuclear weapons delivery vehicles or other means.

46. "Deployment of a nuclear weapon" means prepare or maintain a

nuclear weapon for possible use by any of the following:

a. placing it on, in or near a delivery system;

b. moving it to or maintaining it at a location suitable for delivery to a target.

47. "Destroy" means, with regard to a nuclear weapon, to remove the warhead from its delivery vehicle, dismantle and irreversibly disable the warhead and its components and dismantle and disable or convert the delivery vehicle to non-nuclear use, in accordance with the provisions of this Convention.

48. "Disable" means:

a. with regard to a nuclear weapon, to render the weapon unable to be detonated by such means as disengaging or removing the arming fuse and firing mechanisms;

b. with regard to a plutonium pit, to render it unable to be used in a nuclear weapon, e.g., by disfiguring, quenching, squeezing, dilution, mixing with highly radioactive waste, immobilization and disposition, transmutation or other means;

c. with regard to command and control systems for nuclear weapons, to render such systems incapable of initiating or directing the launch of nuclear weapons delivery vehicles;

d. with regard to a nuclear weapons delivery vehicle, to render such vehicle unable to launch a nuclear weapon including such means as removing essential components and removing the delivery vehicle from the launch facilities.

49. "Disassemble" or "Dismantle" means:

a. with regard to nuclear weapons, to take apart the warhead and remove the subassemblies, components, and individual parts;

b. with regard to a nuclear weapons delivery vehicle, to separate the essentialcomponent parts, such as warheads, propulsion and guidance units.

50. "Immobilization" means the process of putting nuclear material into non-weapons usable form without irradiation, e.g., by mixing with highly radioactive isotopes and encasing into a matrix of another material in order to render separation of the nuclear material from the matrix technically difficult. Immobilization includes vitrification and encasing nuclear material in ceramic.

51. "Nuclear Disarmament Research" means research intended to further the purposes of this Convention.

52. "Nuclear Weapons Research" means experimental or theoretical work undertaken principally to acquire new knowledge going beyond

publicly available information of phenomena and observable facts directed toward understanding, development, improvement, testing, production, deployment, or use of nuclear weapons.

53. "Nuclear Weapons Testing" means nuclear explosions, computer simulations, hydrodynamic tests, hydronuclear tests designed to simulate behavior of nuclear materials, nuclear warheads, nuclear weapons or their components, under nuclear explosive conditions, and subcritical testing using nuclear materials.

54. "Reprocessing" means the separation of irradiated nuclear material and fission products in whole or in part.

55. "Threat of Use of Nuclear Weapons" means any act, whether physical or verbal, including the maintenance of a previously stated policy that creates or is intended to create a perception that a nuclear weapon may or will be used.

56. "Uranium Enrichment" means the process of increasing the percentage of U-235 isotopes so that the uranium can be used as reactor fuel or in nuclear weapons.

57. "Use of Nuclear Weapons" means the detonation of a nuclear weapon.

G. Verification

58. "Verification" means a comprehensive system for ensuring the compliance with and implementation of this Convention. Verification measures include obtaining, providing, and assuring the accuracy of information on nuclear weapons, nuclear material, nuclear facilities, and nuclear weapons delivery vehicles, including information in archives, data bases, and transportation systems, through declarations, monitoring, agreements on sharing information, consultation and clarification, on-site inspections, confidence-building measures, reporting and protection, preventive controls, and any other measures deemed necessary by the Agency.

59. "Abuse of the Right of Verification" means obtaining information, or attempting to obtain information, through verification activities, for purposes not relating to the verification or implementation of and compliance with this Convention.

60. "Confidence-Building Measures" means voluntary measures by States Parties to supply information, additional to that required, to the Technical Secretariat or to other States Parties in order to develop greater confidence in compliance with the Convention. These could

include bilateral or multilateral agreements on monitoring and information sharing between States Parties.

61. "Dual-access" means access to nuclear weapons, nuclear material, or nuclear facilities that requires authorization of a State Party and another State Party or the Agency.

62. "Reconstruction" means undertaking good faith scientifically sound efforts to produce or reproduce data that is not readily available regarding past production of nuclear material. Reconstruction measures include gathering and reviewing past data records, analyzing production capacity and estimating the range of quantity of nuclear material produced, and interviewing individuals with knowledge of the operation of a nuclear facility under review.

63. "Preventive Controls" mean provisions adopted by the Agency to ensure that nuclear material and nuclear facilities are not used for any military or other purpose prohibited under this Convention.

a. The goals of preventive controls include:

i. Timely detection of diversion of nuclear material to allow a response before the material can be fabricated into a nuclear weapon;

ii. Deterring clandestine activities through the possibility of detection;

iii. Prevention of diversion through physical safety procedures and transfer of national access to dual-access.

b. Preventive controls encompass safeguards of the IAEA (including all provisions of the 93+2 Programme), EURATOM, ABACC or other bodies; agreements among States; and agreements between States and the Agency.

c. Preventive controls apply to all nuclear weapons, nuclear material and nuclear facilities. The degree of restrictiveness, accountability and accessibility vary according to the risks posed by these weapons, materials or facilities to the purposes of this Convention. Preventive controls may include:

i. Accountancy and surveillance of nuclear material in any form;

ii. Containment of special nuclear material in any form;

iii. Guidelines for the transport, treatment, handling, storage and disposition of nuclear material;

iv. Environmental guidelines;

v. Dual-access agreements for all nuclear weapons facilities and nuclear storage facilities for proscribed nuclear material.

64. "Technical Means" means the independent gathering or analysis of information which may have relevance to verification of the Convention, without physically accessing the territory being inspected.

65. "National Technical Means" (NTM) comprise nationally-owned and -operated technologies and techniques used to monitor the activities of other states, including their compliance with treaty obligations. [NTM include satellites, aircraft, remote monitoring, signals intelligence (SIGINT) and open source information.]

66. "Open Skies" means a regime for the conduct of observation flights by States Parties over the territories of other States Parties.

H. Delivery Vehicles

67. "Nuclear Weapons Delivery Vehicle" means any vehicle designed for or capable of delivering a nuclear weapon. Any nuclear weapon delivery vehicle which has been constructed, developed, flight-tested or deployed for weapon delivery shall be considered a nuclear weapon delivery vehicle.

68. "Ballistic Missile" means a missile that

a. consists of single or multiple stage(s) whose sole means of propulsion is an internal rocket engine that is functional over all or a portion of the flight path;

b. follows a ballistic trajectory over the remaining unpowered portion of a flight path; and

c. is devoid of active aerodynamic surfaces.

69. "Air-to-Surface Ballistic Missile (ASBM)" means a ballistic missile that is installed in an aircraft or on its external mountings for the purpose of being launched from this aircraft.

70. "Ground-Launched Ballistic Missile (GLBM)" means a ground-launched ballistic missile that is a weapon-delivery vehicle.

71. "Intercontinental Ballistic Missile (ICBM)" means a land-based ballistic missile with a range in excess of 5,500 kilometers.

72. "Submarine [Sea] Launched Ballistic Missile (SLBM)" means a ballistic missile designed to be launched from a submarine or other naval vessel.

73. "Cruise Missile" means an unmanned, self-propelled weapon delivery vehicle that sustains flight through the use of aerodynamic lift over most of its flight path. Cruise Missiles include:

a. Air Launched Cruise Missile (ALCM);

b. Ground Launched Cruise Missile (GLCM);

c. Sea Launched Cruise Missile (SLCM).

74. "Intermediate-Range Missile" means a ballistic or cruise missile having a range capability in excess of 1,000 kilometers but not in excess of 5,500 kilometers.

75. "Shorter-Range Missile" means a ballistic or cruise missile having a range capability equal to or in excess of 500 kilometers but not in excess of 1,000 kilometers.

76. "Bomber" means an airplane which was initially constructed or later converted to be equipped for bombs or air-to-surface missiles.

77. "Heavy Bomber" means a bomber which satisfies either of the following criteria: a. its range is greater than 8,000 kilometers; or b. it is equipped for long-range nuclear ALCMs.

78. "Nuclear-Capable" in relation to delivery vehicles means able to deliver and activate a nuclear weapon.

79. "Nuclear-Capable Missile" means a missile able to deliver any payload over 300 kilometers.

80. "Nuclear-Capable Submarines" includes ballistic missile submarines, cruise missile submarines and attack submarines capable of delivery of nuclear weapons.

III. Declarations

A. Nuclear Weapons

Each State Party shall submit to the Registry, not later than [30] days after this Convention enters into force for it, the following declarations, in which it shall, in accordance with the standards and guidelines set forth in the Verification Annex:

1. Declare whether it owns or possesses or has owned or possessed any nuclear weapons, or whether there are any nuclear weapons located in any place under its jurisdiction or control.

2. Specify the precise location, aggregate quantity and detailed inventory of nuclear weapons it owns or possesses, or that are located in any place under its jurisdiction or control.

3. Report any nuclear weapons on its territory that are owned or possessed by another State or under the jurisdiction or control of another State, whether or not that State is a Party to this Convention.

4. Declare whether it has transferred or received, directly or indirectly, nuclear weapons and specify the transfer or receipt of such weapons.

5. Provide its general plan for destruction of nuclear weapons that it owns or possesses, or that are located in any place under its jurisdiction or control.

B. Nuclear Material

Each State Party shall submit to the Registry the following declarations, in which it shall, in accordance with the standards and guidelines set forth in the Verification Annex:

6. Not later than [60] days after this Convention enters into force for it, declare an inventory of all special nuclear material it owns or possesses or that is located within its jurisdiction or control, whether intended for civilian or military use.

7. Not later than [90] days after this Convention enters into force for it, declare an inventory of all other nuclear material it owns or possesses or that is located within its jurisdiction or control, whether intended for civilian or military use.

8. Not later than [120] days after this Convention enters into force for it, submit a report on the availability of data with respect to nuclear material produced in the past, including estimates regarding missing data and extent of uncertainty, and its plans for the reconstruction of such data.

C. Nuclear Facilities

Each State Party shall submit to the Registry, not later than [180] days after this Convention enters into force for it, the following declarations, in which it shall, in accordance with the standards and guidelines set forth in the Verification Annex:

9. With respect to nuclear weapons facilities:

a. Declare whether it has or has had any nuclear weapons facility under its ownership or possession, or that is or has been located in any place under its jurisdiction or control at any time.

b. Declare any nuclear weapons facility it has or has had under its ownership or possession or that is or has been located in any place under its jurisdiction or control at any time.

c. Declare any nuclear weapons facility on its territory that another State has or has had under its ownership or possession and that is or has been located in any place under the jurisdiction or control of another State at any time.

d. Declare the precise location and production and storage capacities of any facility reported under subparagraphs a, b, or c above.

e. Declare whether it has transferred or received, directly or indirectly, any equipment for the production of nuclear weapons, and provide a detailed account thereof.

f. Specify actions to be taken for the closure of any facility reported under subparagraphs a, b, or c above.

g. Provide its general plan for conversion of any facility reported under subparagraphs a, b, or c into a nuclear weapons destruction facility.

10. With respect to other nuclear facilities, declare the precise location, nature and scope of activities of any nuclear facility under its ownership or possession, or located in any place under its jurisdiction or control. Such declaration shall include, inter alia, laboratories and test and evaluation sites as well as any other facility, site, or installation in which nuclear activities of any kind have been or are carried out, or which are suitable for carrying out such activities.

D. Delivery Vehicles

Each State Party shall submit to the Registry, not later than [210] days after this Convention enters into force for it, the following declarations, in which it shall, in accordance with the standards and guidelines set forth in the Verification Annex:

11. Declare the number and location of all nuclear-capable ballistic and cruise missiles, including all those in production, storage or under repair.

12. Declare the number and location of all nuclear-capable submarines, naval crafts, and aircraft, including all those in production, storage or under repair.

IV. Phases for Implementation

A. General Requirements

1. Each phase indicates the deadline for completion of specific implementation activity. Any phase can begin at any time, and does not require the completion of previous phases before initiation.

2. Implementation activities shall be conducted in accordance with the Verification Annex.

B. Extension of Deadlines

3. If a State Party is unable to complete any of its obligations under Phase One within the deadline, it may submit a request to the Executive Council for an extension. Such a request must be made at least [four] months prior to the deadline, and no extension may exceed [six] months.

4. If a State Party is unable to complete any of its obligations under Phase Two within the deadline, it may submit a request to the Executive Council for an extension. Such a request must be made at least [six] months prior to the deadline, and no extension may exceed [one] year[s].

5. If a State Party is unable to complete any of its obligations under Phases Three, Four, or Five within the deadlines, it may submit a request to the Executive Council for an extension of the deadline. Such a request must be made at least [one] year[s] prior to the deadline for that phase, and no extension may exceed [one] year[s].

C. Reciprocity in Extensions

6. If any State Party makes a request for an extension of any deadline, any other State Party may request a similar extension within [one month] of the original State's request.

D. Phases

7. Phase One. Not later than [one year] after entry into force of this Convention:

  a. All States Parties shall have complied with the requirements of Article III {Declarations}.

  b. Targeting coordinates and navigational information for all nuclear weapons delivery vehicles shall be removed.

  c. All nuclear weapons and nuclear weapons delivery vehicles shall be disabled and dealerted.

  d. Activities listed in Schedule 1 of the Annex on Nuclear Activities shall have ceased.

  e. Production of nuclear weapon components and equipment listed in Schedules 1 and 2 of the Annex on Nuclear Weapons Components and Equipment shall have ceased.

  f. All nuclear weapons testing facilities, nuclear weapons research facilities and nuclear weapons production facilities shall be designated for decommissioning and closure or for conversion.

g. Production of proscribed nuclear material shall have ceased, with the exception of exemption quantities.

h. [Funding for] nuclear weapons research of any sort not consistent with the purposes and obligations of this Convention shall have ceased.

i. Plans for the implementation of all obligations under this Convention shall have been submitted to the Agency.

8. Phase Two. Not later than [two] years after entry into force of this Convention:

a. All nuclear weapons and nuclear weapons delivery vehicles shall be removed from deployment sites.

b. All warheads shall be removed from their delivery vehicles and either placed into nuclear weapons storage facilities or dismantled.

c. Agreements shall be negotiated to subject all nuclear weapons, nuclear material and nuclear facilities to preventive controls.

9. Phase Three. Not later than [five] years after entry into force of this Convention:

a. All nuclear weapons shall be dismantled.

b. All nuclear weapons shall be destroyed, except:

i. no more than [1000] warheads in each of the stockpiles of Russia and the United States; and

ii. no more than [100] warheads in each of the stockpiles of China, France, and the United Kingdom.

c. All nuclear weapons delivery vehicles shall be destroyed or converted for purposes not prohibited under this Convention.

d. All nuclear weapons facilities shall be designated for decommissioning and closure or for conversion.

10. Phase Four. Not later than [10] years after entry into force of this Convention:

a. All nuclear weapons shall be destroyed, except:

i. no more than [50] warheads in each of the stockpiles of Russia and the United States, and

ii. no more than [10] warheads in each of the stockpiles of China, France, and the United Kingdom.

b. All reactors using highly enriched uranium shall be closed or converted to low enriched uranium use.

c. [All reactors using plutonium as fuel shall be closed or converted to reactors that do not use any special nuclear material.]

d. All special nuclear material in any form shall be under strict, effective and exclusive preventive controls.

11. Phase Five. Not later than [15] years after entry into force of this Convention:

a. All nuclear weapons shall be destroyed.

b. [The powers and functions of the Agency shall be reviewed and adjusted to preserve its role in carrying out the objectives of this Convention.]

E. Special Provision

12. The Executive Council may make special provision for temporary retention of small and diminishing quantities of nuclear weapons or proscribed nuclear materials by Nuclear Capable States.

13. States meeting the criteria of this Special Provision shall follow the requirements, guidelines and phases outlined in this Article. They shall not be expected to implement the provisions of this Convention in advance of other States Parties, nor shall they be exempted from the requirements of each phase.

V. Verification

A. Elements of Verification Regime

In order to verify compliance with this Convention, a verification regime shall be established consisting of the following elements:

1. Agreements on sharing data and verification activities among States, UN organs and with existing agencies,

2. A Registry,

3. An International Monitoring System,

4. Reporting of information gathered by National Technical Means,

5. Open Skies,

6. Preventive controls,

7. Consultation and clarification,

8. On-site inspections, including challenge inspections,

9. Confidence-building measures, including additional voluntary measures,

10. Citizen and non-governmental reporting reporting and protection,

11. Any other measures deemed necessary by the Agency.

B. Activities, Facilities, and Materials Subject to Verification

12. All obligations of States Parties and persons as defined, inter alia, in Article I {General Obligations}, Article III {Declarations} and Article IV, Section D {Phases} shall be subject to verification in accordance with the relevant provisions of this Convention and its Verification Annex.

C. Rights and Obligations of States Parties with Respect to Verification

13. Verification activities shall be based on objective information, shall be limited to the subject matter of this Convention, and shall be carried out on the basis of full respect for the sovereignty of States Parties and in the least intrusive manner possible consistent with the effective and timely accomplishment of their objectives. Each State Party shall refrain from any abuse of the right of verification.

14. Each State Party undertakes in accordance with this Convention to cooperate through its National Authority established pursuant to Article VI {National Implementation Measures} of this Convention, with the Agency, with other States Parties and with other agencies as stipulated in this Convention and in separate agreements to facilitate the verification of compliance with this Convention by, inter alia:

a. Establishing the necessary facilities, or providing necessary modifications to existing facilities, to participate in these verification measures, and establishing the necessary communication;

b. Providing all relevant data obtained by technical means and by national systems that are part of the International Monitoring System as agreed among States;

c. Participating, as necessary, in a consultation and clarification process;

d. Permitting the conduct of on-site inspections;

e. Participating in confidence-building measures; and

f. To the extent possible, internationalizing elements of its National Technical Means and incorporating them into the International Monitoring System.

15. Each State Party shall have the right to take measures not contrary to the provisions of this Convention to prevent disclosure of confidential information and data not related to this Convention.

16. Subject to paragraph 15, information obtained by the Agency

through the verification regime established by this Convention shall be made available to all States Parties in accordance with the relevant provisions of this Convention.

17. The provisions of this Convention shall not be interpreted as restricting the international exchange of data for scientific purposes not prohibited by this Convention.

18. Each State Party undertakes to cooperate with the Agency and with other States Parties in the improvement of the verification regime and in the examination of additional monitoring technologies. Such measures shall, when agreed, be incorporated in amendments to this Convention or changes to the Annexes or, where appropriate, be reflected in the operational manuals of the Technical Secretariat.

D. Confidence-Building Measures

19. Each State Party undertakes to cooperate with the Agency and with other States Parties in implementing various measures additional to those explicitly required under this Convention in order to:

a. Develop greater confidence regarding compliance with the obligations under this Convention, and

b. Assist in the compilation of detailed information by the International Monitoring System.

E. Relation to Other Verification Arrangements

20. The Technical Secretariat may enter into cooperative verification arrangements in accordance with the provisions of Article XIV {Cooperation, Compliance and Dispute Settlement} para. 3 and the provisions of Article XVIII, Section A {Relation to Other International Agreements} para. 2.

21. Nothing in this Convention shall be interpreted as in any way limiting or detracting from the verification arrangements assumed by either State under the Treaties Between the United States of America and the Russian Federation on Reduction and Limitation of Strategic Offensive Arms and the Treaty Between the United States of America and the Russian Federation on the Elimination of Their Intermediate-Range and Shorter-Range Missiles {INF}.

22. Nothing in this Convention shall be interpreted as in any way limiting or detracting from the verification arrangements assumed by Argentina and Brazil under the Agreement on the Exclusively Peaceful

Use of Nuclear Energy.

23. Nothing in this Convention shall be interpreted as in any way limiting or detracting from the verification arrangements, assumed by any State under the Comprehensive Nuclear Test Ban Treaty, or under safeguards agreements and additional protocol agreements with the International Atomic Energy Agency [or under the Fissile Materials Cut-Off Treaty].

F. Implementation

24. Prior to entry into force of this Convention, nothing shall preclude any signatory State from implementing, individually or in agreement with other States, the verification measures of this Convention which are applicable to them. Such measures may include public declarations as detailed in Article III {Declarations}, negotiations with other States for the purposes of verifying bilateral or multilateral reductions of nuclear weapons, and the verification of plans for the destruction of nuclear weapons, disposition of special nuclear material, and destruction or conversion of nuclear weapons facilities or nuclear weapons delivery vehicles.

25. Verification measures adopted pursuant to paragraph 23 may include the formation of a provisional authority for the purpose of overseeing verification activities, including assistance in the development of national implementation plans pursuant to Article VI {National Implementation Measures} of this Convention.

VI. National Implementation Measures

A. Legislative Implementation

1. Each State Party shall, in accordance with its constitutional processes, adopt the necessary legislative measures to implement its obligations under this Convention. In particular, it shall:

   a. Extend its penal legislation to provide, in accordance with Article VII, Section A, for the trial, extradition and punishment of persons who commit crimes as defined in Article I, Section B.

   b. Provide all necessary protection for persons who report violations of this Convention, in accordance with Article VII, Section C.

2. Each State Party shall cooperate with other States Parties in affording

legal assistance toward fulfilling the obligations under paragraph 1.

3. Each State Party, in the implementation of its obligations under this Convention, shall assign the highest priority to ensuring the safety of people and to protecting the environment, and shall cooperate as appropriate with other States Parties in this regard.

B. Relations Between the State Party and the Agency

4. In order to fulfill its obligations under this Convention, each State Party shall designate or establish a National Authority to serve as the national focal point for effective liaison with the Agency and other States Parties. Each State Party shall notify the Agency of its National Authority at the time that this Convention enters into force for it. The responsibilities of the National Authority include:

    a. The preparations and submission of declarations in the registry;

    b. The enactment of new legislation or the revision of existing legislation to facilitate the enforcement of the Convention;

    c. Preparations for receiving inspections, including, inter alia, approval of the list of inspectors, issuing of multiple entry visas for inspectors, providing aircraft clearances, and designating points of entry and exit.

5. Each State Party shall inform the Agency of the legislative and administrative measures taken to implement this Convention.

6. Each State Party undertakes to cooperate with the Agency in the exercise of all its functions and in particular to provide assistance to the Technical Secretariat. This includes cooperation in carrying out any investigation which the Agency may initiate, and to provide or support assistance with investigations of non-complying State Parties and with Parties exposed to danger as a result of violation of this Convention.

7. Each State Party shall disseminate information regarding the requirements of this Convention and shall ensure the inclusion of such information in the training of relevant personnel regarding obligations under this Convention.

8. Each State Party shall transmit relevant information gathered by its National Technical Means to the International Monitoring System.

C. Confidentiality

9. Each State Party shall treat as confidential and afford special

handling to information and data that it receives in confidence from the Agency. Information subject to confidentiality shall include data used for purposes not prohibited under this Convention and state and military technology for dual use vehicles, components and computers.

D. Relation to implementation measures assumed or required under other arrangements

10. Nothing in this Convention shall be interpreted as in any way limiting or detracting from the National Implementation Measures assumed or required by States under the Comprehensive Test Ban Treaty, International Atomic Energy Agency Safeguards agreements, International Convention for the Suppression of Acts of Nuclear Terrorism, [Fissile Materials Cut-Off Treaty] and United Nations Security Council resolution 1540.

VII. Rights and Obligations of Persons

A. Criminal Procedure

1. Any person accused of committing a crime under this Convention within the jurisdiction of a State Party of which such person is a citizen or resident shall be
    a. tried according to the legal process of such State if found within such State, or
    b. surrendered to the International Criminal Court if the crime alleged is within the jurisdiction of such Court and the State concerned is unable or unwilling to undertake adequate criminal procedures.
2. If found within another State Party, such person shall be
    a. tried within such State, or
    b. extradited to the State within the jurisdiction of which the crime is alleged to have been committed, or
    c. surrendered to the International Criminal Court if the crime alleged is within the jurisdiction of such Court and the States concerned are unable or unwilling to undertake adequate criminal procedures.
3. Any person accused of a crime under this Convention shall be assumed to be innocent until proven guilty and have the right to a fair trial and humane treatment, as prescribed by the International Covenant on Civil and Political Rights and other conventions and agreements which have acquired the status of customary international

B. Responsibility to Report Violations

4. Persons shall report any violations of this Convention to the Agency. This responsibility takes precedence over any obligation not to disclose information which may exist under national security laws or employment contracts.

5. [Information received by the Agency under the preceding paragraph shall be held in confidence until formal charges are lodged, except to the extent necessary for investigative purposes.]

C. Protection for Persons Providing Information

Intra-state protection

6. Any person reporting a suspected violation of this Convention, either by a person or a State, shall be guaranteed full civil and political rights including the right to liberty and security of person.

7. States Parties shall take all necessary steps to ensure that no person reporting a suspected violation of this Convention shall have any rights diminished or privileges withdrawn as a result.

8. Any individual who [in good faith] provides the Agency or a National Authority with information regarding a known or suspected violation of this Convention cannot be arrested, prosecuted or tried on account thereof.

9. It shall be an unlawful employment practice for an employer to discriminate against any employee or applicant for employment because such person has opposed any practice as a suspected violation of this Convention, reported such violation to the Agency or a National Authority, or testified, assisted, or participated in any manner in an investigation or proceeding under this Convention.

10. Any person against whom a national decision is rendered on account of information furnished by such person to the Agency about a suspected violation of this Convention may appeal such decision to the Agency within [..] months of being notified of such decision. The decision of the Agency in the matter shall be final.

Inter-State Protection

11. Any person reporting a violation of this Convention to the Agency shall be afforded protection by the Agency and by all States Parties, including, in the case of natural persons, the right of asylum in all other States Parties if their safety or security is endangered in the State Party

in which they permanently or temporarily reside.

Additional Provisions

12. [The Executive Council may decide to award monetary compensation to persons providing important information to the Agency concerning violations of this Convention.]

13. Any person who voluntarily admits to the Agency having committed a violation of this Convention, prior to the receipt by the Agency of information concerning such violation from another source, may be exempt from punishment. In deciding whether to grant such exemption, the Agency shall consider the gravity of the violation involved as well as whether its consequences have not yet occurred or can be reversed as a result of the admission made.

VIII. Agency

A. General Provisions

1. The States Parties to this Convention hereby establish the Agency for the Prohibition of Nuclear Weapons (hereinafter "the Agency") to achieve the object and purpose of this Convention, to ensure the implementation of its provisions, including those for international verification of compliance with it, and to provide a forum for consultation and cooperation among States Parties.

2. All States Parties to this Convention shall be members of the Agency. A State Party shall not be deprived of its membership in the Agency.

3. The seat of the Headquarters of the Agency shall be _____.

4. The organs of the Agency are the Conference of the States Parties, the Executive Council, and the Technical Secretariat. The Technical Secretariat shall oversee the Registry and the International Monitoring System.

5. The Agency shall conduct its verification activities provided for under this Convention in the least intrusive manner possible consistent with the timely and efficient accomplishment of their objectives. It shall request only the information and data necessary to fulfill its responsibilities under this Convention. It shall take every precaution to protect the confidentiality of information on civil and military activities and facilities coming to its knowledge in the implementation of this Convention.

6. In undertaking its verification activities the Agency shall consider measures to make use of advances in science and technology.

7. The costs of the Agency's activities shall be paid by States Parties in accordance with Article XVI {Financing}. The budget of the Agency shall comprise two separate chapters, one relating to administrative and other costs, and one relating to verification costs.

8. A member of the Agency which is in arrears in the payments of its financial contribution to the Agency shall have no vote in the Agency if the amount of its arrears equals or exceeds the amount of the contribution due from it for the preceding two full years. The Conference of the States Parties may, nevertheless, permit such a member to vote if it is satisfied that the failure to pay is due to conditions beyond the control of the member.

B. Conference of the States Parties

Composition, procedures and decision-making

9. The Conference of the States Parties (hereinafter "the Conference") shall be composed of all members of this Agency. Each member shall have one representative in the Conference, who may be accompanied by alternates and advisors.

10. The first session of the conference shall be convened by the depositary not later than 30 days after the entry into force of this Convention.

11. The Conference shall meet in regular sessions which shall be held annually unless it decides otherwise.

12. Special sessions of the Conference shall be convened:

    a. When decided by the Conference;

    b. When requested by the Executive Council;

    c. When requested by any member and supported by one third of the members;

    d. In accordance with paragraph 22 to undertake reviews of the operation of this Convention.

    Except in the case of subparagraph (d) the special session shall be convened not later than 30 days after receipt of the request by the Director-General of the Technical Secretariat, unless specified otherwise in the request.

13. The Conference shall also be convened in the form of an Amendment Conference in accordance with Article XVII {Amendments}.

14. Sessions of the Conference shall take place at the seat of the Agency unless the Conference decides otherwise.

15. The Conference shall adopt its rules of procedure. At the beginning

of each regular session, it shall elect its Chairperson and such other officers as may be required. They shall hold office until a new Chairperson and other officers are elected at the next regular session.

16. A majority of the members of the Agency shall constitute a quorum for the Conference.

17. Each member of the Agency shall have one vote in the Conference

18. The Conference shall take decisions on questions of procedure by a simple majority of the members present and voting. Decisions on matters of substance should be taken as far as possible by consensus. If consensus is not attainable when an issue comes up for decision, the Chairperson shall defer any vote for 24 hours and during this period of deferment shall make every effort to facilitate achievement of consensus, and shall report to the Conference before the end of this period. If consensus is not possible at the end of 24 hours, the Conference shall take the decision by a two-thirds majority of members present and voting unless specified otherwise in this Convention. When the issue arises as to whether the question is one of substance or not, the question shall be treated as a matter of substance unless otherwise decided by the Conference by the majority required for decisions on matters of substance.

Powers and functions

19. The Conference shall be the principal organ of the Agency. It shall consider any questions, matters or issues within the scope of this Convention, including those relating to the powers and functions of the Executive Council and the Technical Secretariat. It may make recommendations and take decisions on any questions, matters or issues related to this Convention raised by a State Party or brought to its attention by the Executive Council.

20. The Conference shall oversee the implementation of this Convention, and act in order to promote its object and purpose. The Conference shall review compliance with this Convention. It shall also oversee the activities of the Executive Council and the Technical Secretariat and may issues guidelines in accordance with this Convention to either of them in the exercise of their functions.

21. The Conference shall:

    a. Consider and adopt at its regular sessions the report, program and budget of the Agency, submitted by the Executive Council, as well as consider other reports;

    b. Decide on the scale of financial contributions to be paid by States

Parties in accordance with paragraph 7;

c. Elect the members of the Executive Council;

d. Appoint the Director-General of the Technical Secretariat (hereinafter referred to as the Director-General");

e. Approve the rules of procedure of the Executive Council submitted by the latter;

f. Establish such subsidiary organs as it finds necessary for the exercise of its functions in accordance with this Convention;.

g. Review scientific and technological developments that could affect the operation of this Convention and, in this context, direct the Director-General to establish a Scientific Advisory Board to enable him or her, in the performance of his or her functions, to render specialized advice in areas of science and technology relevant to this Convention, to the Conference, the Executive Council or States Parties. The Scientific Advisory Board shall be composed of independent experts appointed in accordance with terms of reference adopted by the Conference;

h. Take the necessary measures to ensure compliance with this Convention and to redress

and remedy any situation which contravenes the provisions of this Convention, in accordance with Article XIV {Cooperation, Compliance and Dispute Settlement}.

22. The Conference shall, not later than one year after the expiration of the fifth and the tenth year after the entry into force of this Convention, and at such other times within that time period as may be decided upon, convene in special sessions to undertake reviews of the operation of this Convention. Such reviews shall take into account any relevant scientific and technological developments. At intervals of five years thereafter, unless otherwise decided upon, further sessions of the Conference shall be convened with the same objective.

C. The Executive Council

Composition, procedure and decision-making

23. The Executive Council shall consist of 44 members. Each State Party shall have the right, in accordance with the principle of rotation, to serve on the Executive Council. The members of the Executive Council shall be elected by the Conference for a term of four years. In order to ensure the effective functioning of this Convention, due regard being paid to equitable geographic distribution, to representation by nuclear-capable states and to the interests of all states to be free

from the threat of nuclear devastation, the Executive Council shall be composed as follows:

a. All Nuclear Weapons States Parties; and

b. Six States Parties from the Middle East and South Asia;

c. Seven States Parties from Latin America and the Caribbean;

d. Six States Parties from Eastern Europe;

e. Seven States Parties from Africa;

f. Six States Parties from among North America and Western Europe;

g. Six States Parties from South East Asia, the Pacific and the Far East;

h. Up to two additional States Parties that have special interest or expertise in implementing the aims of this Convention to be elected if required.

24. For the first election of the Executive Council 21 members shall be elected for a term of two

years, and 21 members for a term of four years.

25. The Conference may, on its motion or upon the request of a majority of the members of the Executive Council, review the composition of the Executive Council taking into account developments related to the principles specified in paragraph 23.

26. The Executive Council shall elaborate its rules of procedure and submit them to the Conference for approval.

27. The Executive Council shall elect its Chairperson from among its members.

28. The Executive Council shall meet for regular sessions. Between regular sessions it shall meet as often as may be required for the fulfillment of its powers and functions.

29. Each member of the Executive Council shall have one vote. Unless otherwise specified in this Convention, the Executive Council shall take decisions on matters of substance by a twothirds majority of all its members. When an issue arises as to whether the question is one of substance or not, that question shall be treated as a matter of substance unless otherwise decided by the Executive Council by the majority required for decisions on matters of substance.

Powers and Functions

30. The Executive Council shall be the executive organ of the Agency. It shall be responsible to the Conference. The Executive council shall carry out the powers and functions entrusted to it under this Convention, as well as those functions delegated to it by the Conference. In so doing,

it shall act in conformity with the recommendations, decisions and guidelines of the Conference and assure their proper and continuous implementation.

31. The Executive Council shall promote the effective implementation of, and compliance with, this Convention. It shall supervise the activities of the Technical Secretariat, cooperate with the National Authority of each State Party and facilitate consultations and cooperation among States Parties at their request.

32. The Executive Council shall:
   a. Consider and submit to the Conference the draft program and budget of the Agency;
   b. Consider and submit to the Conference the draft report of the Agency on the implementation of this Convention, the report on the performance of its own activities and such special reports as it deems necessary or which the Conference may request;
   c. Make arrangements for the sessions of the Conference including the preparation of the draft agenda.

33. The Executive Council may request the convening of a special session of the Conference.

34. The Executive Council shall:
   a. Conclude agreements or arrangements with States and international organizations on behalf of the Agency, subject to prior approval by the Conference;
   b. Approve agreements or arrangements relating to the implementation of verification activities, negotiated by the Technical Secretariat with States Parties.

35. The Executive Council shall consider any issue or matter within its competence affecting this Convention and its implementation, including concerns regarding compliance, and cases of non-compliance, and, as appropriate, inform States Parties and request compliance within a specified time.

36. If the Executive Council considers further action to be necessary, it shall take, inter alia, one or more of the following measures in accordance with Article XIV {Cooperation, Compliance and Dispute Settlement}:
   a. Inform all States Parties of the issue or matter;
   b. Bring the issue or matter to the attention of the Conference;
   c. Make recommendations to the Conference regarding measures to redress the situation and to ensure compliance.
   d. The Executive Council shall, in cases of particular gravity and urgency, bring the issue or matter, including relevant information and conclusions, directly to the attention of the United Nations General Assembly and the United Nations Security Council. It shall at the same time inform all States Parties of this step.

D. The Technical Secretariat

37. The Technical Secretariat shall assist the Conference and the Executive Council in the performance of their functions. The Technical Secretariat shall carry out the verification measures provided for in this Convention. It shall carry out the other functions entrusted to it under this Convention as well as those functions delegated to it by the Conference and the Executive Council.

38. With respect to the verification of and compliance with this Convention, the Technical Secretariat shall:
   a. Maintain the Registry and other information databases in accordance with Section F below;
   b. Maintain and coordinate the operation of the International Monitoring System;
   c. Provide technical assistance in, and support for, the installation and operation of monitoring systems;
   d. Assist the Executive Council in facilitating consultation and clarification among States Parties;
   e. Receive requests for on-site inspections and process them, facilitate the Executive Council consideration of such requests, carry out the preparation for, and provide technical support during, the conduct of on-site inspections, and report to the Executive Council;
   f. Negotiate agreements or arrangements relating to the implementation of verification activities with States Parties, subject to approval by the Executive Council;
   g. Provide technical assistance and technical evaluation to States Parties in the implementation of the provisions of this Convention;
   h. Assist the States Parties through their National Authorities on other issues of verification under this Convention.

39. The Technical Secretariat shall develop and maintain, subject to approval by the Executive Council, operational manuals to guide the operation of various components of the verification regime, in accordance with the Verification Annex. These manuals shall not constitute integral parts of this Convention or the Annexes, and may be changed by the Technical Secretariat subject to approval by the Executive Council. The Technical Secretariat shall promptly inform the States Parties of any changes in the operational manuals.

40. With respect to administrative matters the Technical Secretariat shall:
   a. Prepare and submit to the Executive Council the draft program and budget of the Agency;
   b. Prepare and submit to the Executive Council the draft report of the Agency on the implementation of this Convention and such

other reports as the Conference or the Executive Council may request;

c. Provide administrative and technical support to the Conference, the Executive Council
and subsidiary organs;

d. Address and receive communications on behalf of the Agency to and from States Parties on matters pertaining to the implementation of this Convention;

e. Upon approval by the Executive Council and the Conference, submit the report of the Agency to the United Nations Secretary-General.

41. All requests and notifications by States Parties to the Agency shall be transmitted through their National Authorities to the Director-General. Requests and notifications shall be in one of the official languages of the United Nations. In response the Director-General shall use the language of the transmitted request or notification.

42. The Technical Secretariat shall inform the Executive Council of any problem that has arisen with regard to the discharge of its functions, including doubts, ambiguities or uncertainties about compliance with this Convention that have come to its notice in the performance of its verification activities or through confidential or non-governmental sources and that it has been unable to resolve or clarify through its consultations with the State Party concerned.

43. The Technical Secretariat shall comprise a Director-General, who shall be its head and chief administrative officer, inspectors and such scientific, technical and other personnel as may be required.

44. The Inspectorate shall be a unit of the Technical Secretariat and shall act under the supervision of the Director-General.

45. The Director-General shall be appointed by the Conference upon the recommendation of the Executive Council for a term of four years, renewable for one further term, but not thereafter. The appointment of the Director-General shall be considered a matter of substance governed by paragraph 18.

46. The Director-General shall be responsible to the Conference and the Executive Council for the appointment of the staff and the organization and functioning of the Technical Secretariat. The paramount consideration in the employment of the staff and in the determination of the conditions of service shall be the necessity of securing the highest standards of efficiency, competence and integrity. Only citizens of States Parties shall serve as the Director-General, as inspectors or as other members of the professional and clerical staff.

Due regard shall be paid to the importance of recruiting the staff on as wide a geographical basis as possible. Recruitment shall be guided by the principle that the staff shall be kept to a minimum necessary for the proper discharge of the responsibilities of the Technical Secretariat.
47. The Director-General shall be responsible for the organization and functioning of the Scientific Advisory Board referred to in paragraph 21.g The Director-General shall, in consultation with States Parties and non-governmental sources, appoint members of the Scientific Advisory Board, who shall serve in their individual capacity. The members of the Board shall be appointed on the basis of their expertise in the particular scientific fields relevant to the implementation of this Convention. The Director-General may also, as appropriate, in consultation with members of the Board, establish temporary working groups of scientific experts to provide recommendations on specific issues. In regard to the above, States Parties and non-governmental sources may submit lists of experts to the Director- General. The Scientific Advisory Board may be called upon to review nuclear or other research and determine whether it is of a nature prohibited under this Convention or of a nature that may contribute to verification of nuclear disarmament.
48. In the performance of their duties, the Director-General, the inspectors and the other members of the staff shall not seek or receive instructions from any Government or from any other source external to the Agency. They shall refrain from any action that might reflect on their positions as international officers responsible only to the Conference and the Executive Council.
49. Each State Party shall respect the exclusively international character of the responsibilities of the Director-General, the inspectors and the other members of the staff and not seek to influence them in the discharge of their responsibilities.

E. Privileges and Immunities

50. The Agency shall enjoy on the territory and in any other place under the jurisdiction or control of a State Party such legal capacity and such privileges and immunities as are appropriate for the exercise of its functions.
51. Delegates of States Parties, together with their alternates and advisers, representatives appointed to the Executive Council together with their alternates and advisers, the Director- General and the

staff of the Agency shall enjoy such privileges and immunities as are necessary in the independent exercise of their functions in connection with the Agency.

52. The legal capacity, privileges, and immunities referred to in this Article shall be defined in agreements between the Agency and the States Parties as well as in an agreement between the Agency and the State in which the headquarters of the Agency is seated.

53. Notwithstanding paragraphs 50 and 51, the privileges and immunities enjoyed by the Director- General and the staff of the Technical Secretariat during the conduct of verification activities shall be those set forth in the Verification Annex.

F. Registry and Other Databases

54. The Technical Secretariat shall maintain a Registry of the following:
    a. All nuclear weapons;
    b. All nuclear material;
    c. All nuclear facilities;
    d. All nuclear weapons delivery vehicles;
    e. Any other facilities or materials as determined by the Technical Secretariat.

55. The Technical Secretariat shall obtain information from the following sources:
    a. Declarations by States in accordance with the provisions of Article III {Declarations};
    b. Reports by States on progress in implementing their obligations under this Convention;
    c. The International Monitoring System;
    d. National Technical Means;
    e. Systematic inspections;
    f. Challenge inspections;
    g. Other organizations with which the Agency has concluded agreements on sharing information in accordance with Article XVIII, Section A {Relation to Other International Agreements};
    h. Other inter-governmental and non-governmental organizations that collect and submit such information;
    i. Publicly available sources;
    j. Any other sources which the Technical Secretariat deems appropriate

53. Notwithstanding paragraphs 50 and 51, the privileges and immunities enjoyed by the Director- General and the staff of the Technical Secretariat during the conduct of verification activities shall be those set forth in the Verification Annex.

F. Registry and Other Databases

54. The Technical Secretariat shall maintain a Registry of the following:
    a. All nuclear weapons;
    b. All nuclear material;
    c. All nuclear facilities;
    d. All nuclear weapons delivery vehicles;
    e. Any other facilities or materials as determined by the Technical Secretariat.

55. The Technical Secretariat shall obtain information from the following sources:
    a. Declarations by States in accordance with the provisions of Article III {Declarations};
    b. Reports by States on progress in implementing their obligations under this Convention;
    c. The International Monitoring System;
    d. National Technical Means;
    e. Systematic inspections;
    f. Challenge inspections;
    g. Other organizations with which the Agency has concluded agreements on sharing information in accordance with Article XVIII, Section A {Relation to Other International Agreements};
    h. Other inter-governmental and non-governmental organizations that collect and submit such information;
    i. Publicly available sources;
    j. Any other sources which the Technical Secretariat deems appropriate.

56. The Technical Secretariat shall make available to the Registry information obtained from the above sources with the exception of information which may remain confidential because of legitimate national and international security concerns or trade secret concerns.

57. Information in the Registry shall be available to all States parties and to the public according to criteria established by separate agreements [among States].

G. International Monitoring System

58. The International Monitoring System shall comprise facilities and systems for monitoring by satellite, on-site sensors, remote sensors, radionuclide sampling, respective means of communication, aircraft and other systems developed as deemed necessary by the Agency.

59. The International Monitoring System shall be placed under the authority of the Technical Secretariat.

60. All monitoring facilities of the International Monitoring System shall be owned and operatedby the States hosting or otherwise taking responsibility for them except for those systems or facilities which may be owned or operated by another agency or by the United Nations, or constructed or acquired by the Agency in accordance with paragraph 64.

61. The Technical Secretariat shall acquire equipment necessary for collating and analyzing data provided by the International Monitoring System.

62. Any State Party may, if it so decides and upon agreement with the Technical Secretariat, give a monitoring facility to the Agency.

63. The Technical Secretariat may, upon agreement of the Conference and in accordance with its funding guidelines, construct or otherwise acquire a monitoring system or facility if it determines that such a facility or system is necessary for verification of obligations of States under this Convention, and if no State is able or willing to provide such a system or facility or information from such a system or facility to the International Monitoring System.

64. Each State shall have the right to participate in the international exchange of data and to have access to all data made available to the Registry.

65. The Agency shall conclude agreements with other agencies or organizations using international monitoring systems relating to the sharing of information obtained through such systems relevant to the verification of this Convention in accordance with Article XVIII, Section A {Relation to Other International Agreements}.

66. Data obtained by the International Monitoring System not directly relevant to verification of this Convention shall be treated as confidential, except where such information is relevant to. the verification of another international agreement [and there is an agreement on sharing such information between the Agency and the organization responsible for implementation of that agreement].

67. Data obtained from the International Monitoring System shall first be analyzed, processed and verified by the Technical Secretariat before being compiled as part of the Registry, in accordance with the provisions of paragraph 57.

IX. Nuclear Weapons

A. General Requirements

1. All nuclear weapons [with corresponding delivery vehicles] shall be

taken off alert status, disabled, removed from deployment, declared, and destroyed in accordance with the guidelines and standards of Article III {Declarations}, Article IV {Phases for Implementation}, the Verification Annex, and the provisions set forth below:

B. Procedures for Destroying Nuclear Weapons

2. Each State Party shall take the following measures with respect to all nuclear weapons that it owns or possesses or that are under its jurisdiction or control:
   a. All warheads shall be bar-coded, registered, and tagged for identification using secure visual tags.
   b. All nuclear weapons shall be destroyed or moved to nuclear weapons storage facilities subject to international preventive controls. No exclusive national access to the repositories is allowed. Weapons may be removed from the nuclear weapons storage facilities only for the purposes of destruction.
   c. All core elements from newly dismantled warheads shall be quenched or otherwise deformed and placed in storage under international preventive controls until final disposal of the proscribed nuclear material, in accordance with the guidelines and standards of Article X {Nuclear Material}.

C. Prevention of Production of Nuclear Weapons

3. All nuclear [weapons] facilities and deployment sites shall be subject to verification, including challenge inspections at any time and non-destructive detection of hidden warheads, to ensure compliance with obligations under this Convention not to develop, produce, or deploy nuclear weapons.

X. Nuclear Material

A. Reconstruction and Documentation

1. All military and civilian nuclear material shall be documented and declared according to the guidelines and standards set forth in Article III {Declarations} and the Verification Annex.
2. Special Nuclear Material
   a. Records of production and use of special nuclear material produced in the past shall be reconstructed to the extent possible through analysis of past records, measures of transparency including

national legislation aimed at disclosure of information, interviews, and any other appropriate means.

b. All special nuclear material storage sites and related nuclear facilities usable for production of special nuclear material shall be subject to preventive controls, including inventory verification as set forth in the Verification Annex.

B. Control of Special Nuclear Material

3. Subject to Section C below, production and use of proscribed nuclear material is prohibited. Existing inventories of special nuclear material shall be subject to preventive controls and

storage and disposal in accordance with the guidelines and standards set forth below and in separate verification agreements.

4. All treatment of nuclear material that improves its quality to the level of proscribed nuclear material or improves the accessibility of proscribed nuclear material is prohibited, including, inter alia, separation of plutonium from spent fuel, enrichment of uranium in U-235 beyond unavoidable civilian requirements or beyond 20%, or extraction of tritium from heavy water, with the exception of exemption quantities.

5. All existing stocks of special nuclear material shall be placed under preventive controls until a safe method of final disposal is found and approved by the Agency. All handling of proscribed nuclear material except for such handling as necessary for the purposes of this Convention shall be prohibited.

6. [Burning of special fissionable material is prohibited unless the net amount of fissionable material resulting from such burning is reduced.]

7. Facilities for the production, research and testing of special nuclear material may be converted to uses consistent with the purposes and obligations of this Convention. Conversion of such

facilities may include research and development for methods of demilitarization and disposal of proscribed nuclear material, including immobilization and final disposition of plutonium.

C. Licensing Requirements

8. The Agency shall establish a licensing process for civilian use of proscribed nuclear material which is not prohibited.

D. Relation to other International Agreements

9. Nothing in this Section shall be interpreted as in any way limiting or detracting from the verification arrangements assumed by any State under safeguards agreements and additional protocol agreements with the International Atomic Energy Agency [or under the Fissile Materials Cut-Off Treaty].

XI. Nuclear Facilities

A. Nuclear Weapons Facilities

1. All nuclear weapons production facilities shall cease operations prohibited under this Convention and shall be closed or converted to purposes not prohibited under this Convention.
2. All nuclear weapons testing facilities shall cease operations and shall be permanently closed [or converted to purposes not prohibited under this Convention].
3. All nuclear weapons research facilities shall be closed or converted to research in accordance with paragraph 4.
4. Funding of research for the purposes of designing, modernizing, constructing, modifying or maintaining reliability of nuclear weapons is prohibited. Funding of research for the purpose of developing knowledge in the physics of nuclear explosions is prohibited. Funding of research in safety mechanisms for existing nuclear weapons is permitted only until all nuclear weapons are dismantled. Funding of research for the purposes of safe dismantling and destroying of nuclear weapons and for safe disposal of special nuclear material is permitted.
5. [All nuclear reprocessing facilities shall cease operations and shall be permanently closed.]
6. All nuclear facilities shall be subject to preventive controls.
7. All plans for the destruction or conversion of nuclear weapons [production, research and testing facilities and principal nuclear] facilities, submitted in accordance with Article IV {Phases for Implementation}, shall include provisions or recommendations for the placement of former employees of such facilities in positions of employment consistent with their experience and expertise and with the object and purpose of this Convention. Such positions and recommendations may include employment within a converted facility, employment for the destruction of a nuclear facility, employment for

the destruction of nuclear weapons or disposition of special nuclear material, or employment within the Agency for the purposes of verification.

B. Command, Control, and Communications Facilities and Deployment Sites

8. Each State Party shall make the following changes to nuclear targeting commands and command systems in accordance with Article IV {Phases for Implementation}:
    a. Rescind alert status on all nuclear weapons;
    b. Remove targeting coordinates from all command and control systems; and
    c. Remove navigational information for all nuclear armed missiles from the navigational systems.
9. Each State Party shall, in accordance with Article IV {Phases for Implementation} and the Verification Annex, destroy any facility, system or sub-system designed or used solely for the
purpose of launching, targeting, directing or detonating a nuclear weapon or its delivery vehicle, or for aiding or assisting in any of these purposes.
10. Each State Party shall, in accordance with Article IV {Phases for Implementation} and the Verification Annex, and in order to prevent use for purposes prohibited under this Convention, destroy or convert any facility, system or sub-system which is used for the purpose of launching, targeting, directing or detonating a nuclear weapon or its delivery vehicle, or for aiding or assisting in any of these purposes, and which is also used for purposes not prohibited under this Convention.
11. Any facility, system or sub-system designed and used for detection of activities prohibited under this Convention is permitted.
12. All plans for the destruction or conversion of command, control, and communications facilities and deployment sites submitted in accordance with Article IV {Phases for Implementation} and the Verification Annex, shall include provisions or recommendations for the placement of former employees of such facilities in positions of employment consistent with their experience and expertise and with the object and purpose of this Convention. Such positions and recommendations may include employment within a converted facility, employment for the destruction of a nuclear facility, employment for the purpose of gathering information, including National Technical Means, and employment within the Agency for the purposes of inspection or other methods of verification.

C. Nuclear reactors, enrichment and reprocessing facilities, nuclear materials storage sitesand other nuclear-fuel cycle locations outside of facilities

13. All States shall declare the precise location, nature and scope of nuclear reactors, enrichment and reprocessing facilities, nuclear laboratories, nuclear materials storage sites and other nuclear-fuel cycle locations outside of facilities.

14. All plutonium reprocessing facilities shall cease operations and be permanently closed.

15. All States shall conclude safeguards agreements with the Agency [or International Atomic Energy Agency] to verify that nuclear facilities are operated consistent with obligations under this convention including obligations under Section X (Nuclear Material).

D. Activities at nuclear facilities

16. Activities undertaken at nuclear facilities that are listed in Schedule 1 of the Annex on Nuclear Activities shall be prohibited.

17. Activities undertaken at nuclear facilities that are listed in Schedule 2 of the Annex on Nuclear Activities are permitted unless otherwise determined by the Conference of States Parties in accordance with Articles XIV (Cooperation, Compliance and Dispute Settlement).

18. Activities undertaken at nuclear facilities that are listed in Schedule 3 of the Annex on Nuclear Activities are permitted.

XII. Nuclear Weapons Delivery Vehicles

1. All deployment, development, testing, production, or acquisition of delivery vehicles and launchers designed solely for the purpose of delivering nuclear weapons {Schedule 1} is prohibited.

2. All delivery vehicles and launchers designed solely for the purpose of delivering nuclear weapons shall be destroyed according to Article IV {Phases for Implementation} and the Verification Annex.

3. All delivery vehicles capable of use for the delivery of nuclear weapons or non-nuclear weapons {Schedule 2} shall be destroyed according to Article IV {Phases for Implementation} or converted for purposes not prohibited under this Convention.

140

Schedule 1 - Nuclear Weapons Delivery Vehicles to Be Destroyed

Intercontinental Ballistic Missiles
Submarine Launched Ballistic Missiles
Heavy Bombers
Ballistic Missile Submarines
Ground Launched Cruise Missile

Schedule 2 - Delivery Vehicles To be Destroyed or Converted

Air-to-Surface Ballistic Missiles
Ground Launched Ballistic Missiles
Air Launched Cruise Missile
Sea Launched Cruise Missile
Nuclear-capable fighter bombers
Cruise Missile Submarines
Attack Submarines
Warships

[Schedule 3 - Transport Vehicles Not Designed for Nuclear Weapons to be Subject to Preventive Controls]

XIII. Activities Not Prohibited Under This Convention

1. Each State Party has the right, subject to the provisions of this Convention [and other agreements and regulations relating to nuclear material] to the research, development and use of nuclear energy for peaceful purposes.
2. Each State Party shall adopt the necessary measures to ensure that research, development and use of nuclear energy within its territory or under its control is undertaken only for purposes not prohibited under this Convention. To this end, and in order to verify that activities are in accordance with obligations under this Convention, each State Party shall subject nuclear facilities and nuclear material listed in the Annex on Nuclear Activities, Components and Equipment of this Convention, or any other activities so declared by the Agency, to control and verification measures as provided in Sections V (Verification), VI (National Implementation Measures), VIII (Agency), X (Nuclear Material), XI (Nuclear Facilities) [and the Verification Annex].
3. Each State Party has the right to the research, development,

production, acquisition and deployment of weapons-delivery systems for security purposes. This right is subject to the provisions of this Convention, other agreements and regulations relating to weapons and weapons systems, the United Nations Charter and other international law relating to the threat or use of force.

4. In the exercise of military activities not prohibited under this Convention,] each State Part shall adopt the necessary measures to ensure that [weapons and] weapons delivery systems are only developed, produced, otherwise acquired, retained, transferred, tested or deployed in a manner consistent with this Convention. To this end, and in order to verify that activities are in accordance with obligations under this Convention, each State Party shall subject weapons delivery systems including command, communication, control and production facilities to control and verification measures as provided in Section XII (Nuclear Weapons Delivery Vehicles) [and the Verification Annex].

XIV. Cooperation, Compliance and Dispute Settlement

A. Consultation, Cooperation, and Fact-finding

1. States Parties shall consult and cooperate, directly among themselves, or through the Agency or other appropriate international procedures, including procedures within the framework of the United Nations and in accordance with its Charter, on any matter which may be raised relating to the object and purpose, or the implementation of the provisions, of this Convention.

2. Each State Party undertakes to cooperate with the Agency and with other States Parties in the improvement of the verification, destruction and conversion regimes, with a view to developing specific measures to enhance the efficient, safe and cost-effective verification, destruction and conversion procedures and methods of this Convention.

3. Without prejudice to the right of any State Party to request a challenge inspection, States Parties should, whenever possible, first make every effort to clarify and resolve, through exchange of information and consultations among themselves, any matter which may cause doubt about compliance with this Convention, or which gives rise to concerns about a related matter which may be considered ambiguous. A State Party which receives a request from another State Party for clarification of any matter which the requesting State Party believes causes such a doubt or concern shall provide the requesting State Party as soon as

possible, but in any case not later than [48] hours after the receipt of a request to clarify a possible threat of use of nuclear weapons or [10] days after the receipt of a request to clarify any other matter, with information sufficient to answer the doubt or concern raised along with an explanation of how the information provided resolves the matter. Nothing in this Convention shall affect the right of any two or more States Parties to arrange by mutual consent for inspections or any other procedures among themselves to clarify and resolve any matter which may cause doubt about compliance or gives rise to a concern about a related matter which may be considered ambiguous. Such arrangements shall not affect the rights and obligations of any State Party under other provisions of this Convention.

Procedure for requesting clarification

4. A State Party shall have the right to request the Executive Council to assist in clarifying any situation which may be considered ambiguous or which gives rise to a concern about the possible non-compliance of another State Party with this Convention. The Executive Council shall provide appropriate information in its possession relevant to such a concern.

5. A State Party shall have the right to request the Executive Council to obtain clarification from another State Party on any situation which may be considered ambiguous or which gives rise to a concern about its possible non-compliance with this Convention. In such a case, the following shall apply:

   a. The Executive Council shall forward the request for clarification to the State Party concerned through the Director-General not later than [24] hours after its receipt;
   b. The requested State Party shall provide the clarification to the Executive Council as soon as possible, but in any case not later than [48] hours after the receipt of a request to clarify possible threat or use of nuclear weapons or [10] days after the receipt of a request to clarify any other matter;
   c. The Executive Council shall take note of the clarification and forward it to the requesting State Party not later than [24] hours after its receipt;
   d. If the requesting State Party deems the clarification to be inadequate, it shall have the right to request the Executive Council to obtain from the requested State Party further clarification;
   e. For the purpose of obtaining further clarification requested under subparagraph d, the Executive Council may call on the Director-General to establish a group of experts from the Technical Secretariat, or if appropriate staff are not available in the Technical Secretariat, from elsewhere, to examine all available information

and data relevant to the situation causing the concern. The group of experts shall submit a factual report to the Executive Council on its findings;

f. If the requesting State Party considers the clarification obtained under subparagraphs d and e to be unsatisfactory, it shall have the right to request a special session of the Executive Council in which States Parties involved that are not members of the Executive Council shall be entitled to take part. In such a special session, the Executive Council shall consider the matter and may recommend any measure it deems appropriate to resolve the situation.

6. A State Party shall also have the right to request the Executive Council to clarify any situation which has been considered ambiguous or has given rise to a concern about its possible noncompliance with this Convention. The Executive Council shall respond by providing such assistance as appropriate.

7. The Executive Council shall inform the States Parties about any request for clarification provided in this Article.

8. If the doubt or concern of a State Party about a possible non-compliance has not been resolved within [60] days after the submission of the request for clarification to the Executive Council, or it believes its doubts warrant urgent consideration, notwithstanding its right to request a challenge inspection, it may request a special session of the Conference in accordance with Article VIII {Agency}. At such a special session, the Conference shall consider the matter and may recommend any measure it deems appropriate to resolve the situation.

Procedures for challenge inspections

9. Each State Party has the right to request an on-site challenge inspection of any facility or location in the territory or in any other place under the jurisdiction or control of any other State Party for the sole purpose of clarifying and resolving any questions concerning possible noncompliance with the provisions of this Convention, and to have this inspection conducted anywhere without delay by an inspection team designated by the Director-General and in accordance with the Verification Annex.

10. Each State Party is under the obligation to keep the inspection request within the scope of this Convention and to provide in the inspection request all appropriate information on the basis of which a concern has arisen regarding possible non-compliance with this Convention as specified in the Verification Annex. Each State Party shall refrain from unfounded inspection requests, care being taken to avoid abuse. The challenge inspection shall be carried out for the sole

purpose of determining facts relating to the possible non-compliance.

11. For the purpose of verifying compliance with the provisions of this Convention, each State Party shall permit the Technical Secretariat to conduct the on-site challenge inspection pursuant to paragraph 9.

12. Pursuant to a request for a challenge inspection of a facility or location, and in accordance with the procedures provided for in the Verification Annex, the inspected State Party shall have:

    a. The right and the obligation to make every reasonable effort to demonstrate its compliance with this Convention and, to this end, to enable the inspection team to fulfil its mandate;

    b. The obligation to provide access within the requested site for the sole purpose of establishing facts relevant to the concern regarding possible non-compliance; and

    c. The right to take measures to protect sensitive installations, and to prevent disclosure of confidential information and data, not related to this Convention.

13. With regard to an observer, the following shall apply:

    a. The requesting State Party may, subject to the agreement of the inspected State Party, send a representative who may be a national either of the requesting State Party or of a third State Party, to observe the conduct of the challenge inspection.

    b. The inspected State Party shall then grant access to the observer in accordance with the Verification Annex.

    c. The inspected State Party shall, as a rule, accept the proposed observer, but if the inspected State Party exercises a refusal, that fact shall be recorded in the final report.

14. The requesting State Party shall present an inspection request for an on-site challenge inspection to the Executive Council and at the same time to the Director-General for immediate processing.

15. The Director-General shall immediately ascertain that the inspection request meets the requirements specified in the Verification Annex, and, if necessary, assist the requesting State Party in filing the inspection request accordingly. When the inspection request fulfills the requirements, preparations for the challenge inspection shall begin.

16. The Director-General shall transmit the inspection request to the inspected State Party not less than 12 hours before the planned arrival of the inspection team at the point of entry.

17. After having received the inspection request, the Executive Council shall take cognizance of the Director-General's actions on the request and shall keep the case under its consideration throughout the inspection procedure. However, its deliberations shall not delay the inspection process.

18. The Executive Council may, not later than 12 hours after having

received the inspection request, decide by a three-quarter majority of all its members against carrying out the challenge inspection, if it considers the inspection request to be frivolous, abusive or clearly beyond the scope of this Convention as described in paragraph 9. Neither the requesting nor the inspected State Party shall participate in such a decision. If the Executive Council decides against the challenge inspection, preparations shall be stopped, no further action on the inspection request shall be taken, and the States Parties concerned shall be informed accordingly.

19. The Director-General shall issue an inspection mandate for the conduct of the challenge inspection. The inspection mandate shall be the inspection request referred to in paragraphs 9 and 10 put into operational terms, and shall conform with the inspection request.

20. The challenge inspection shall be conducted in accordance with the provisions of the Verification Annex. The inspection team shall be guided by the principle of conducting the challenge inspection in the least intrusive manner possible, consistent with the effective and timely accomplishment of its mission.

21. The inspected State Party shall assist the inspection team throughout the challenge inspection and facilitate its task. If the inspected State Party proposes, pursuant to the Verification Annex, arrangements to demonstrate compliance with this Convention, alternative to full and comprehensive access, it shall make every reasonable effort, through consultations with the inspection team, to reach agreement on the modalities for establishing the facts with the aim of demonstrating its compliance.

22. The final report shall contain the factual findings as well as an assessment by the inspection team of the degree and nature of access and cooperation granted for the satisfactory implementation of the challenge inspection. The Director-General shall promptly transmit the final report of the inspection team to the requesting State Party, to the inspected State Party, to the Executive Council and to all other States Parties. The Director-General shall further transmit promptly to the Executive Council the assessments of the requesting and of the inspected States Parties, as well as the views of other States Parties which may be conveyed to the Director-General for that purpose, and then provide them to all States Parties.

23. The Executive Council shall, in accordance with its powers and functions, review the finalreport of the inspection team as soon as it is presented, and address any concerns as to:

a. Whether any non-compliance has occurred;
b. Whether the request had been within the scope of this Convention; and
c. Whether the right to request a challenge inspection had been abused.

24. If the Executive Council reaches the conclusion, in keeping with its powers and functions, that further action may be necessary with regard to paragraph 23, it shall take the appropriate measures to redress the situation and to ensure compliance with this Convention, including specific recommendations to the Conference. In the case of abuse, the Executive Council shall examine whether the requesting State Party should bear any of the financial implications of the challenge inspection.

25. The requesting State Party and the inspected State Party shall have the right to participate in the review process. The Executive Council shall inform the States Parties and the next session of the Conference of the outcome of the process.

26. If the Executive Council has made specific recommendations to the Conference, the Conference shall consider action in accordance with Section B.

B. Measures to Redress a Situation and to Ensure Compliance, Including Sanctions

27. The Conference, taking into account the recommendations of the Executive Council, shall take necessary measures, as set forth in paragraphs 28, 29 and 30 to ensure compliance with this Convention and to redress and remedy any situation which contravenes the provisions of this Convention.

28. In cases where a State Party has been requested by the Conference or the Executive Council to redress a situation raising problems with regard to its compliance and fails to fulfill the request within the specified time, the Conference may, inter alia, decide to restrict or suspend the State Party from the exercise of its rights and privileges under this Convention until the Conference decides otherwise.

29. In cases where damage to the object and purpose of this Convention may result from noncompliance with the basic obligations of this Convention, the Conference may recommend to States Parties collective measures which are in conformity with international law. Suchmeasures may include restrictions or suspensions of all assistance in nuclear activities outlined in Schedule 2 of the Annex on Nuclear

Activities, Components and Equipment. If the State concerned continues in its failure to comply with the request, further sanctions may be imposed.

30. The Conference, or alternatively, if the case is urgent, the Executive Council, may bring the issue, including relevant information, conclusions and recommendations, to the attention of the United Nations General Assembly and the United Nations Security Council.

31. The threat or use of nuclear weapons shall be deemed to be a threat to the peace subject to the provisions of the United Nations Charter.

C. Settlement of Disputes

32. Disputes that may arise concerning the application, implementation or interpretation of this Convention shall be settled in accordance with the relevant provisions of this Convention, including Section B and in conformity with the provisions of the Charter of the United Nations.

33. When a disputes arises between two or more States Parties, or between one or more States Parties and the Agency, relating to the application, implementation or interpretation of this Convention, the parties concerned shall consult together with a view to the expeditious settlement of the dispute by negotiation, mediation, arbitration or by other peaceful means of the parties' choice, including recourse to appropriate organs of this Convention and, by mutual consent, referral to the International Court of Justice in conformity with the Statute of the Court.

34. If other peaceful means of settlement are not found, a State Party in dispute with one or more States Parties may refer the dispute to the International Court of Justice, in conformity with the Statute of the Court [and the Optional Protocol Concerning the Compulsory Settlement of Disputes]. The States Parties involved shall keep the Executive Council informed of actions being taken.

35. The Executive Council may contribute to the settlement of a dispute by whatever means it deems appropriate, including offering its good offices, calling upon the States Parties to a dispute to start the settlement process of their choice and recommending a time-limit for any agreed procedure.

36. The Conference shall consider questions related to disputes raised by States Parties or brought to its attention by the Executive Council. The Conference shall, as it finds necessary, establish or entrust organs with tasks related to the settlement of these disputes in conformity

with Article VIII {Agency}.

37. The Conference and the Executive Council may recommend to the General Assembly of the United Nations, to request the International Court of Justice to give an advisory opinion on any legal question arising within the scope of the activities of the Agency. An agreement between the Agency and the United Nations shall be concluded for this purpose in accordance with Article VIII {Agency}.

38. This Section is without prejudice to Sections A and B.

XV. Entry Into Force

A. Conditions of Entry Into Force

1. This Convention shall enter into force [180] days after the date on which the following conditions are met:
   a. [All] Nuclear Weapons States have deposited their instruments of ratification; and
   b. All Nuclear Capable States have deposited their instruments of ratification; and
   c. At least [65] States in total have deposited instruments of ratification [including at least [40] States from Annex IV: List of Countries with Nuclear Power Reactors], [or] [including at least [40] States from Annex V: List of Countries with Nuclear Power Reactors or Nuclear Research Reactors].

2. For States whose instruments of ratification or accession are deposited subsequent to the entry into force of this Convention, it shall enter into force on the 30th day following the date of deposit of their instrument of ratification or accession.

B. State Waiver of Entry Into Force Requirements

For States who waive the entry into force requirements, this Convention shall enter into force on the 30th day following the date of deposit of their instrument of ratification or accession.

XVI. Financing

1. The costs of the Agency's activities shall be paid by States Parties in accordance with the United Nations scale of assessment adjusted to take into account differences in membership between the United Nations and this Agency. The budget of the Agency shall comprise two separate chapters, one relating to administrative and other costs, and

one relating to verification and compliance costs.

2. Each Nuclear Weapons State shall meet the costs of destruction of weapons, proscribed nuclear material and nuclear facilities under its authority. Each Nuclear Weapons State shall meet the costs of verification of nuclear facilities under its authority, except for instances of challenge inspections which are funded according to the provisions of the Verification Annex.

3. The Agency shall establish a voluntary fund to assist States Parties to comply with paragraph 2 where such compliance imposes undue financial burdens on them.

XVII. Amendments

1. Any State Party may propose amendments to this Convention. Any State Party may also propose changes, as specified in paragraph 4, to the Annexes of this Convention. Proposals for amendments shall be subject to the procedures in paragraphs 2 and 3. Proposals for changes, as specified in paragraph 4, shall be subject to the procedures in paragraph 5.

2. The text of a proposed amendment shall be submitted to the Director-General for circulation to

all States Parties and to the Depositary. The proposed amendment shall be considered only by an Amendment Conference. Such an Amendment Conference shall be convened if one third or more of the States Parties notify the Director-General [not later than [60 days] after its circulation] that they support further consideration of the proposal. The Amendment Conference shall be held immediately following a regular session of the Conference unless the requesting States Parties ask for an earlier meeting. In no case shall an Amendment Conference be held less than 60 days after the circulation of the proposed amendment.

3. Amendments shall enter into force for all States Parties 20 days after deposit of the instruments of ratification or acceptance by all the States Parties referred to under subparagraph b below:
    a. When adopted by the Amendment Conference by a positive vote of a majority of all States Parties [with no State Party casting a negative vote]; and
    b. Ratified or accepted by all those States Parties casting a positive vote at the Amendment Conference.

4. In order to ensure the viability and the effectiveness of this Convention, provisions in the Annexes shall be subject to changes in

accordance with paragraph 5, if proposed changes are related only to matters of an administrative or technical nature.

5. Proposed changes referred to in paragraph 4 shall be made in accordance with the following procedures:

a. The text of the proposed changes shall be transmitted together with the necessary information to the Director-General. Additional information for the evaluation of the proposal may be provided by any State Party and the Director-General. The Director- General shall promptly communicate any such proposals and information to all States Parties, the Executive Council and the Depositary;

b. Not later than 60 days after its receipt, the Director-General shall evaluate the proposal to determine all its possible consequences for the provisions of this Convention and its implementation and shall communicate any such information to all States Parties and the Executive Council;

c. The Executive Council shall examine the proposal in the light of all information available to it, including whether the proposal fulfills the requirements of paragraph 4. Not later than 90 days after its receipt, the Executive Council shall notify its recommendation, with appropriate explanations, to all States Parties for consideration. States Parties shall acknowledge receipt within 10 days;

d. If the Executive Council recommends to all States Parties that the proposal be adopted, it shall be considered approved if no State Party objects to it within 90 days after receipt of the recommendation. If the Executive Council recommends that the proposal be rejected, it shall be considered rejected if no State Party objects to the rejection within 90 days after receipt of the recommendation;

e. If a recommendation of the Executive Council does not meet with the acceptance required under subparagraph d, a decision on the proposal, including whether it fulfills the requirements of paragraph 4, shall be taken as a matter of substance by the Conference at its next session;

f. The Director-General shall notify all States Parties and the Depositary of any decision under this paragraph;

g. Changes approved under this procedure shall enter into force for all States Parties 180 days after the date of notification by the Director-General of their approval unless another time period is recommended by the Executive Council or decided by theConference.

XVIII. Scope and Application of Convention

A. Relation to other International Agreements

1. Nothing in this Convention shall be interpreted as in any way limiting or detracting from the obligations assumed by any State under the United Nations Charter; the Treaty on the Non- Proliferation of Nuclear Weapons; the Treaty Banning Nuclear Weapon Tests in the Atmosphere, in Outer Space and Under Water; the Treaty for the Prohibition of Nuclear Weapons in Latin America and the Caribbean; the Treaty on the Prohibition of the Emplacement of Nuclear Weapons and Other Weapons of Mass Destruction on the Sea-Bed and the Ocean Floor and in the Subsoil Thereof; the Agreement Governing the Activities of States on the Moon and Other Celestial Bodies; the South Pacific Nuclear Free Zone Treaty; the African Nuclear Free Zone Treaty; the Southeast Asia Nuclear Weapon Free Zone Treaty; the Central Asia Nuclear Weapon Free Zone Treaty; any other treaties establishing nuclear weapon free zones; the Comprehensive Nuclear Test Ban Treaty; the Treaty Between the U.S.A. and the U.S.S.R. on the Elimination of Their Intermediate-Range and Short- Range Missiles; the Treaty Between the U.S.A. and the U.S.S.R. on the Reduction and Limitation of Strategic Offensive Arms; the Treaty Between the U.S.A. and Russia on Further Reduction and Limitation of Strategic Offensive Arms; the Treaty between Russia and the United States on Strategic Offensive Reductions, the International Convention for the Suppression of Acts of Nuclear Terrorism, or under agreements with the International Atomic Energy Agency.

2. Pursuant to Article VIII {Agency}, the Agency may enter into agreements with the implementing organizations of other international agreements for the purpose of sharing information necessary or applicable to the verification tasks of each organization involved, or for any other purposes that would further the objectives of the international agreements concerned.

B. Status of the Annexes

3. The Annexes form an integral part of this Convention. Any reference to this Convention includes the Annexes.

C. Duration and Withdrawal

4. This Convention shall be of unlimited duration.
5. Withdrawal from this Convention shall not be permitted [upon ratification by all Nuclear Weapons States].

D. Reservations

6. The Articles of this Convention shall not be subject to reservations. The Annexes of this Convention shall not be subject to reservations incompatible with its object and purpose.

XIX. Conclusion of Convention

A. Signature

1. This Convention shall be open for signature for all States before its entry into force.

B. Ratification

2. This Convention shall be subject to ratification by States Signatories according to their respective constitutional processes.

C. Accession

3. Any State which does not sign this Convention before its entry into force may accede to it at any time thereafter.

D. Depository

4. The Secretary-General of the United Nations is hereby designated as the Depository of this Convention and shall, inter alia:
a. Promptly inform all signatory and acceding States of the date of each signature, the date of deposit of each instrument of ratification or accession and the date of the entry into force of this Convention, and of the receipt of other notices;
b. Transmit duly certified copies of this Convention to the Governments of all signatory and acceding States; and
c. Register this Convention pursuant to Article 102 of the Charter of the United Nations.

E. Authentic Texts

5. This Convention, of which the Arabic, Chinese, English, French, Russian and Spanish texts are equally authentic, shall be deposited with the Secretary-General of the United Nations.

Optional Protocol Concerning the Compulsory Settlement of Disputes

The States Parties to this Protocol, expressing their wish to resort to the compulsory jurisdiction of the International Court of Justice, unless some other form of settlement is provided for in the Convention or has been agreed upon by the Parties within a reasonable period, have agreed as follows:
Disputes arising out of the interpretation or application of this Convention shall lie within the compulsory jurisdiction of the International Court of Justice, and may accordingly be brought before the Court by an application by any party to the dispute being a Party to this Protocol.

Optional Protocol Concerning Energy Assistance

The States Parties to this Protocol:
Desiring to prevent any threat to the aims and objectives of this Convention from arising due to the proliferation of nuclear technology which could aid or assist in the development of nuclear weapons,
Desiring further to prevent any threat to health and the environment arising from the excessive creation of radionuclides in nuclear reactors,
Affirming the right to the development of sustainable and environmentally safe energy sources,
Have agreed as follows:
1. Not to manufacture, assemble, transfer or otherwise acquire nuclear power reactors.
2. Not to use any existing power reactor, nor the products from the use of any nuclear power reactor.
3. To close any existing nuclear power reactors within [five] years of signing this protocol.
4. To assist other Parties to this protocol in the development and use of non-nuclear, sustainable energy sources.
5. To create a voluntary fund for the purposes of implementing paragraph 4.

Annex I. Nuclear Activities

A. Guidelines for Schedules of Nuclear Activities

Guidelines for Schedule 1

1. The following criteria shall be taken into account in considering whether a nuclear activity shall be included in Schedule 1:
    (a) It is an activity specifically prohibited under Article I of this Convention.
    (b) It is an activity the purpose of which is to aid or assist in any activity specifically prohibited under Article I of this Convention.
    (c) It is an activity which poses a grave risk to the object and purpose of this Convention by virtue of its high potential for aiding and assisting activities specifically prohibited by this Convention.
    (d) It has little or no use for purposes not prohibited under this Convention, or alternatively its use for purposes not prohibited under this Convention can be safely substituted by another activity.

2. Schedule 1 activities are prohibited.

Guidelines for Schedule 2

3. The following criteria shall be taken into account in considering whether a nuclear activity shall be included in Schedule 2:
    (a) It is an activity not specifically prohibited under Article I of this Convention.
    (b) It is an activity the purpose of which is not to aid or assist in any activity specifically prohibited under Article I of this Convention.
    (c) It is an activity which poses some risk to the object and purpose of this Convention by virtue of its potential to aid and assist activities specifically prohibited by this Convention.

4. Schedule 2 activities are permitted unless otherwise determined by the Conference in accordance with Articles [Agency, Technical Secretariat] and [compliance].

Guidelines for Schedule 3

5. The following criteria shall be taken into account in considering whether a nuclear activity shall be included in Schedule 3:

(a) It is an activity not specifically prohibited under Article I of this Convention.

(b) It is an activity the purpose of which is not to aid or assist in any activity specifically prohibited under Article I of this Convention.

(c) It is an activity which poses no risk to the object and purpose of this Convention.

6. Schedule 3 activities are permitted.

B. Schedules of Nuclear Activities
Schedule 1

(1) Production of nuclear weapons

(2) Use of nuclear weapons

(3) Threat of use of nuclear weapons

(4) Production and any use of special nuclear material

(5) Production of metals or alloys containing plutonium or uranium

(6) Weaponization: This covers the research, development, manufacturing and testing required to make nuclear explosive devices from special fissionable or fusionable material

(7) Nuclear fuel fabrication using plutonium, uranium-233, uranium enriched to 20% or more in uranium-235

(8) Import, construction or use of research and power reactors of any kind utilizing uranium enriched to 20% or more in uranium-235, uranium-233, plutonium or MOX as a fuel or any reactor designed specifically for plutonium production. This includes critical and sub-critical assemblies

(9) Reprocessing of irradiated fuel or irradiation targets containing nuclear-weapons capable material. This includes the use of hot cells and associated equipmen

(10) Enrichment of uranium in isotope U-235 beyond 20% and any preparatory steps in this process, including the preparation and storage of UCl4 and UF6 enriched to more than 3% in U-235. {The preparation of UC14 and UF6 from natural uranium will not be forbidden by the NWC. After enrichment it should not be stored in this form which would be appropriate feeding material for further enrichment beyond 20%.}

(11) Production, separation, and enrichment of the isotope of plutonium-239 , hydrogen, tritium and lithium-6

(12) Production of antiprotons, antimatter, nuclear isomers and super-heavy elements in significant quantities

Schedule 2

(1) Import, construction, use of research and power reactors of any type using natural uranium or uranium enriched to less than 20% in uranium-235 as a fuel. This includes critical and subcritical assemblies, but excludes reactors specifically designed for plutonium production

(2) Prospecting, mining or processing of ores containing uranium and/

or thorium

(3) Preparation of chemical compounds containing uranium enriched to less than 20% in uranium-235 and thorium; excluding the preparation of UCI4 and UF6 enriched to more than 3% in U-235

(4) Nuclear fuel fabrication using natural uranium or uranium enriched to less than 20% in uranium-235

(5) Production of particle and laser beams of all kind

(6) Nuclear fusion experimental devices based on inertial confinement, including diagnostics

Schedule 3

(1) Application of radiation and isotopes in food and agriculture:
- soil fertility, irrigation and crop production
- [plant breeding and genetics]
- animal production and health
- insect and pest control
- [food preservation]
- other uses upon approval

(2) Applications of radiation and isotopes in medicine
- diagnostic and therapeutic medicine including dosimetry
- Radiotherapy by teletherapy and brachytherapy
- nutrition and health-related environmental studies
- other uses upon approval

(3) Application of radiation and isotopes in industrial processes
- Radiography and other non-destructive testing methods
- Industrial process control and quality control
- Radiotracer applications in oil, chemical and metallurgical processes
- Development of water and mineral resources
- Industrial radiation processing
- Other uses upon approval

(4) Applications in research with and production and disposal of radioactive isotopes and elementary particles
- Conditioning and disposal of radioactive wastes
- Nuclear fusion experimental devices based on magnetic confinement, including diagnostics
- Production of isotopes both radioactive and stable. The production of the isotope Pu-239, titanium and lithium-6 is prohibited.
- Import, construction and use of neutron sources, electron accelerators, particle accelerators, heavy ion accelerators
- Research on radiation physics and chemistry and on the physical and chemical properties of isotopes except in areas relevant to activities not prohibited by or subject to authorization under this Convention

Annex II. Nuclear Weapon Components

Guidelines for Schedule 1

1. A component shall be included in Schedule 1 if it is produced solely for the purpose of incorporation into a nuclear explosive device.
2. Manufacture, transfer or stockpiling of Schedule 1 components is prohibited.

Guidelines for Schedule 2

3. The following criteria shall be taken into account in considering whether a component shall be included in Schedule 2:
   (a) The component is produced for incorporation into a nuclear explosive device.
   (b) The component is also used for purposes not prohibited under this convention, but is not produced in large commercial quantities for such purposes.
   (c) There exist alternative components for the purposes cited in paragraph (b).
4. Manufacture, transfer or stockpiling of Schedule 2 components is prohibited.

Guidelines for Schedule 3

5. The following criteria shall be taken into account in considering whether a component shall be included in Schedule 3:
   (a) The component is produced for incorporation into a nuclear explosive device.
   (b) The component is also used for purposes not prohibited under this convention, but is not produced in large commercial quantities for such purposes.
   (c) There do not exist alternative components for the purposes cited in paragraph (b).
6. Manufacture, transfer or stockpiling of Schedule 3 components is permitted only in accordance with the provisions established by the Agency.

Guidelines for Schedule 4

7. The following criteria shall be taken into account in considering whether a component shall be included in Schedule 4:

(a) The component is produced for incorporation into a nuclear explosive device.

(b) The component is also used for purposes not prohibited under this convention, and is produced in large commercial quantities for such purposes.

[(c) There do not exist alternative components for the purposes cited in paragraph (b).]

8. Manufacture of Schedule 4 components is permitted only in accordance with the provisions established by the Agency.

Annex III. List of countries and geographical regions for the purpose of Article VIII.C.23

Africa

Algeria, Angola, Benin, Botswana, Burkina Faso, Burundi, Cameroon, Cape Verde, Central African Republic, Chad, Comoros, Congo, Cote d'Ivoire, Djibouti, Egypt, Equatorial Guinea, Eritrea, Ethiopia, Gabon, Gambia, Ghana, Guinea, Guinea-Bissau, Kenya, Lesotho, Liberia, Libyan Arab Jamahiriya, Madagascar, Malawi, Mali, Mauritania, Mauritius, Morocco, Mozambique, Namibia, Niger, Nigeria, Rwanda, Sao Tome & Principe, Senegal, Seychelles, Sierra Leone, Somalia, South Africa, Sudan, Swaziland, Togo, Tunisia, Uganda, United Republic of Tanzania, Zaire, Zambia, Zimbabwe.

Eastern Europe

Albania, Armenia, Azerbaijan, Belarus, Bosnia and Herzegovina, Bulgaria, Croatia, Czech Republic, Estonia, Georgia, Hungary, Latvia, Lithuania, Moldova, Poland, Romania, Russian Federation, Slovakia, Slovenia, the Former Yugoslav Republic of Macedonia, Ukraine, Yugoslavia.

Latin America and the Caribbean

Antigua and Barbuda, Argentina, Bahamas, Barbados, Belize, Bolivia, Brazil, Chile, Colombia, Costa Rica, Cuba, Dominica, Dominican Republic, Ecuador, El Salvador, Grenada, Guatemala, Guyana, Haiti, Honduras, Jamaica, Mexico, Nicaragua, Panama, Paraguay, Peru, Saint Kitts and Nevis, Saint Lucia, Saint Vincent and the Grenadines, Suriname, Trinidad and Tobago, Uruguay, Venezuela.

Middle East and South Asia

Afghanistan, Bahrain, Bangladesh, Bhutan, India, Iran (Islamic Republic of), Iraq, Israel, Jordan, Kazakhstan, Kuwait, Kyrgyzstan, Lebanon, Maldives, Nepal, Oman, Pakistan, Qatar, Saudi Arabia, Sri Lanka, Syrian Arab Republic, Tajikistan, Turkmenistan, United Arab Emirates, Uzbekistan, Yemen.

North America and Western Europe

Andorra, Austria, Belgium, Canada, Cyprus, Denmark, Finland, France, Germany, Greece, Holy See, Iceland, Ireland, Italy, Liechtenstein, Luxembourg, Malta, Monaco, Netherlands, Norway, Portugal, San Marino, Spain, Sweden, Switzerland, Turkey, United Kingdom of Great Britain and Northern Ireland, United States of America.

South East Asia, the Pacific and the Far East

Australia, Brunei Darussalam, Cambodia, China, Cook Islands, Democratic People's Republic of Korea, Fiji, Indonesia, Japan, Kiribati, Lao People's Democratic Republic, Malaysia, Marshall Islands, Micronesia (Federated States of), Mongolia, Myanmar, Nauru, New Zealand, Niue, Palau, Papua New Guinea, Philippines, Republic of Korea, Samoa, Singapore, Solomon Islands, Thailand, Tonga, Tuvalu, Vanuatu, Viet Nam.

Annex IV. List of countries with nuclear power reactors

ARGENTINA, ARMENIA, BELGIUM, BRAZIL, BULGARIA, CANADA, CHINA, FINLAND, FRANCE, GERMANY, HUNGARY, INDIA, IRAN, JAPAN, REPUBLIC OF KOREA, LITHUANIA, MEXICO, NETHERLANDS, PAKISTAN, ROMANIA, RUSSIA, S. AFRICA, SLOVAKIA, SLOVENIA, SPAIN, SWEDEN, SWITZERLAND, UNITED KINGDOM, UKRAINE, USA

Annex V. List of countries and geographical regions with nuclear power reactors and/or nuclear research reactors

ARGENTINA, ARMENIA, AUSTRALIA, AUSTRIA, BANGLADESH, BELARUS, BELGIUM, BRAZIL, BULGARIA

CANADA,, CHILE, CHINA, COLOMBIA, CZECH REPUBLIC, DPRK, DEMOCRATIC REPUBLICOF CONGO, DENMARK, EGYPT, EUROPEAN UNION, FINLAND, FRANCE, GEORGIA, GERMANY, GHANA, GREECE,, HUNGARY, INDIA, INDONESIA, IRAN, IRAQ, ISRAEL, ITALY, JAMAICA, JAPAN, KAZAKHSTAN, LATVIA, LIBYA, LITHUANIA, MALAYSIA, MEXICO, MOROCCO, NETHERLANDS, NIGERIA, NORWAY, PAKISTAN, PERU PHILIPPINES,, POLAND, PORTUGAL, REPUBLIC OF KOREA, ROMANIA, RUSSIA, SERBIA AND MONTENEGRO, SLOVAKIA, SLOVENIA, SOUTH AFRICA, SPAIN, SWEDEN, SWITZERLAND, SYRIA, TAIWAN, THAILAND, TUNISIA, TURKEY, UNITED KINGDOM, UKRAINE, USA

New Zealand Nuclear Free Zone, Disarmament, and Arms Control Act 1987

New Zealand Nuclear Free Zone, Disarmament, and Arms Control Act 1987

Public Act 1987 No 86
Date of assent 8 June 1987
Commencement 8 June 1987[1]

1 Short Title

This Act may be cited as the New Zealand Nuclear Free Zone, Disarmament, and Arms Control Act 1987.

2 Interpretation

In this Act, unless the context otherwise requires,—
biological weapon means any agent, toxin, weapon, equipment, or means of delivery referred to in Article 1 of the Convention on the Prohibition of the Development, Production and Stockpiling of Bacteriological (Biological) and Toxin Weapons and on their Destruction of 10 April 1972 (the text of which is set out in Schedule 5)
foreign military aircraft means any aircraft, as defined in section 2 of the Defence Act 1971, which is for the time being engaged in the service of or subject to the authority or direction of the military authorities of any State other than New Zealand
foreign warship means any ship, as defined in section 2 of the Defence Act1971, which—
(a) belongs to the armed forces of a State other than New Zealand; and
(b) bears the external marks that distinguishes ships of that State's nationality; and
(c) is under the command of an officer duly commissioned by the

1   This Appendix version of the Act has been formatted editorially for the purposes of this book. For an original version of the Act see: file:///C:/Users/gdarn/OneDrive/Books/TPNW/References/New%20Zealand%20Nuclear%20Free%20Zone%20Disarmament%20and%20Arms%20Control%20Act%201987.pdf.

Government of that State; and

(d) is manned by a crew under regular armed forces discipline

immunities, in relation to any ship, aircraft, or crew member, means immunities enjoyed under international law by ships, aircraft, or crew members of a class to which that ship, aircraft, or crew member belongs

internal waters of New Zealand means the internal waters of New Zealand as defined by section 4 of the Territorial Sea, Contiguous Zone, and Exclusive Economic Zone Act 1977

nuclear explosive device means any nuclear weapon or other explosive device capable of releasing nuclear energy, irrespective of the purpose for which it could be used, whether assembled, partly assembled, or unassembled; but does not include the means of transport or delivery of such a weapon or device if separable from and not an indivisible part of it

passage means continuous and expeditious navigation without stopping or anchoring except in as much as these are incidental to ordinary navigation or are rendered necessary by distress or for the purpose of rendering assistance to persons, ships, or aircraft in distress

territorial sea of New Zealand means the territorial sea of New Zealand as defined by section 3 of the Territorial Sea, Contiguous Zone, and Exclusive Economic Zone Act 1977.

Section 2 **internal waters of New Zealand**: amended, on 1 August 1996, pursuant to section 5(4) of the Territorial Sea and Exclusive Economic Zone Amendment Act 1996 (1996 No 74).

Section 2 **territorial sea of New Zealand**: amended, on 1 August 1996, pursuant to section 5(4) of the Territorial Sea and Exclusive Economic Zone Amendment Act 1996 (1996 No 74).

3 Act to bind the Crown

This Act shall bind the Crown.

4 New Zealand Nuclear Free Zone

There is hereby established the New Zealand Nuclear Free Zone, which shall comprise:

(a) all of the land, territory, and inland waters within the territorial limits of New Zealand; and

(b) the internal waters of New Zealand; and

(c) the territorial sea of New Zealand; and

(d) the airspace above the areas specified in paragraphs (a) to (c).

Prohibitions in relation to nuclear explosive devices and biological weapons

5 Prohibition on acquisition of nuclear explosive devices

(1) No person, who is a New Zealand citizen or a person ordinarily resident in New Zealand, shall, within the New Zealand Nuclear Free Zone,—
  (a) manufacture, acquire, or possess, or have control over, any nuclear explosive device; or
  (b) aid, abet, or procure any person to manufacture, acquire, possess, or have control over any nuclear explosive device.
(2) No person, who is a New Zealand citizen or a person ordinarily resident in New Zealand, and who is a servant or agent of the Crown, shall, beyond the New Zealand Nuclear Free Zone,—
  (a) manufacture, acquire, or possess, or have control over, any nuclear explosive device; or
  (b) aid, abet, or procure any person to manufacture, acquire, possess, or have control over any nuclear explosive device.

6 Prohibition on stationing of nuclear explosive devices

No person shall emplant, emplace, transport on land or inland waters or internal waters, stockpile, store, install, or deploy any nuclear explosive device in the New Zealand Nuclear Free Zone.

7 Prohibition on testing of nuclear explosive devices

No person shall test any nuclear explosive device in the New Zealand Nuclear Free Zone.

8 Prohibition of biological weapons

No person shall manufacture, station, acquire, or possess, or have control over any biological weapon in the New Zealand Nuclear Free Zone.

9 Entry into internal waters of New Zealand

(1) When the Prime Minister is considering whether to grant approval to the entry of foreign warships into the internal waters of New Zealand, the Prime Minister shall have regard to all relevant information and advice that may be available to the Prime Minister including information and advice concerning the strategic and security interests of New Zealand.

(2) The Prime Minister may only grant approval for the entry into the internal waters of New Zealand by foreign warships if the Prime Minister is satisfied that the warships will not be carrying any nuclear explosive device upon their entry into the internal waters of New Zealand.

10 Landing in New Zealand

(1) When the Prime Minister is considering whether to grant approval to the landing in New Zealand of foreign military aircraft, the Prime Minister shall have regard to all relevant information and advice that may be available to the Prime Minister including information and advice concerning the strategic and security interests of New Zealand.
(2) The Prime Minister may only grant approval to the landing in New Zealand by any foreign military aircraft if the Prime Minister is satisfied that the foreign military aircraft will not be carrying any nuclear explosive device when it lands in New Zealand.
(3) Any such approval may relate to a category or class of foreign military aircraft, including foreign military aircraft that are being used to provide logistic support for a research programme in Antarctica, and may be given for such period as is specified in the approval.

11 Visits by nuclear powered ships

Entry into the internal waters of New Zealand by any ship whose propulsion is wholly or partly dependent on nuclear power is prohibited.

Savings

12 Passage through territorial sea and straits

Nothing in this Act shall apply to or be interpreted as limiting the freedom of—
   (a) any ship exercising the right of innocent passage (in accordance with international law) through the territorial sea of New Zealand; or
   (b) any ship or aircraft exercising the right of transit passage (in accordance with international law) through or over any strait used for international navigation; or
   (c) any ship or aircraft in distress.

13 Immunities

Nothing in this Act shall be interpreted as limiting the immunities of—
(a) any foreign warship or other government ship operated for non-commercial purposes; or
(b) any foreign military aircraft; or
(c) members of the crew of any ship or aircraft to which paragraph (a) or paragraph (b) applies.

Offences

14 Offences and penalties

(1) Every person commits an offence against this Act who contravenes or fails to comply with any provision of sections 5 to 8.
(2) Every person who commits an offence against this Act is liable on conviction to imprisonment for a term not exceeding 10 years. Section 14(2): amended, on 1 July 2013, by section 413 of the Criminal Procedure Act 2011 (2011 No 81).

15 Consent of Attorney-General to proceedings in relation to offences

(1) No charging document may be filed against any person for—
(a) an offence against this Act; or
(b) the offence of conspiring to commit an offence against this Act; or
(c) the offence of attempting to commit an offence against this Act,—
except with the consent of the Attorney-General:
provided that a person alleged to have committed any offence mentioned in this subsection may be arrested, or a warrant for any such person's arrest may be issued and executed, and any such person may be remanded in custody or on bail, notwithstanding that the consent of the Attorney-General to the filing of a charging document for the offence has not been obtained, but no further or other proceedings shall be taken until that consent has been obtained.
(2) The Attorney-General may, before deciding whether or not to give consent under subsection (1), make such inquiries as the Attorney-General thinks fit. Section 15(1): amended, on 1 July 2013, by section 413 of the Criminal Procedure Act 2011 (2011 No 81).

16 Establishment of Public Advisory Committee on Disarmament and Arms Control

There is hereby established a committee to be called the Public Advisory Committee on Disarmament and Arms Control.

17 Functions and powers of Committee

(1) The functions of the Committee shall be—
  (a) to advise the Minister of Foreign Affairs and Trade on such aspects of disarmament and arms control matters as it thinks fit:
  (b) to advise the Prime Minister on the implementation of this Act:
  (c) to publish from time to time public reports in relation to disarmament and arms control matters and on the implementation of this Act:
  (d) to make such recommendations as it thinks fit for the granting of money from such fund or funds as may be established for the purpose of promoting greater public understanding of disarmament and arms control matters.
(2) The Committee shall have all such powers as are reasonably necessary or expedient to enable it to carry out its functions. Section 17(1)(a): amended, on 1 July 1993, by section 6(1) of the Foreign Affairs Amendment Act 1993 (1993 No 48).

18 Membership of Committee

(1) The Committee shall consist of 9 members, of whom—
  (a) one shall be the Minister for Disarmament and Arms Control, who shall be the Chairman; and
  (b) 8 shall be appointed by the Minister of Foreign Affairs and Trade.
(2) Each member of the Committee appointed under subsection (1) (b) shall be appointed for such term not exceeding 3 years as may be specified in the instrument of appointment, but may from time to time be reappointed.
(3) Any such member may be removed from office for incapacity, neglect of duty, or misconduct proved to the satisfaction of the Minister of Foreign Affairs and Trade, or may resign by notice in writing to that Minister.
(4) The functions and powers of the Committee shall not be affected by any vacancy in its membership.
Section 18(1)(b): amended, on 1 July 1993, by section 6(1) of the Foreign Affairs Amendment Act 1993 (1993 No 48).
Section 18(3): amended, on 1 July 1993, by section 6(1) of the Foreign Affairs Amendment Act 1993 (1993 No 48).

19 Procedure of Committee

Subject to any directives given by the Minister of Foreign Affairs and Trade, the Committee may regulate its procedure in such manner as it thinks fit. Section 19: amended, on 1 July 1993, by section 6(1) of the Foreign Affairs Amendment Act 1993 (1993 No 48).

20 Remuneration and travelling expenses

(1) The Committee is hereby declared to be a statutory Board within the meaning of the Fees and Travelling Allowances Act 1951.

(2) There shall be paid to the members of the Committee, out of money appropriated by Parliament for the purpose, remuneration by way of fees or allowances, and travelling allowances and expenses, in accordance with the Fees and Travelling Allowances Act 1951, and the provisions of that Act shall apply accordingly.

21 Money to be appropriated by Parliament for purposes of this Act

All fees, salaries, allowances, and other expenditure payable or incurred under or in the administration of this Act shall be payable out of money to be appropriated by Parliament for the purpose.

Amendments to Marine Pollution Act 1974
[Repealed]
Heading: repealed, on 20 August 1998, pursuant to section 481(1) of the Maritime Transport Act 1994 (1994 No 104).

22 Interpretation

[Repealed]
Section 22: repealed, on 20 August 1998, by section 481(1) of the Maritime Transport Act 1994 (1994 No 104).

23 Application of Part 2 of Marine Pollution Act 1974

[Repealed]
Section 23: repealed, on 20 August 1998, by section 481(1) of the Maritime Transport Act 1994 (1994 No 104).

24 New sections inserted

[Repealed]
Section 24: repealed, on 20 August 1998, by section 481(1) of the Maritime Transport Act 1994(1994 No 104).

25 Permits

[Repealed]
Section 25: repealed, on 20 August 1998, by section 481(1) of the Maritime Transport Act 1994 (1994 No 104).

Amendments to other Acts

26 Amendment to Diplomatic Privileges and Immunities Act 1968

Amendment(s) incorporated in the Act(s).

27 Amendment to Official Information Act 1982

Amendment(s) incorporated in the Act(s).

28 Amendment to Foreign Affairs Act 1983

[Repealed]
Section 28: repealed, on 1 December 1988, by section 14(1) of the Foreign Affairs Act 1988 (1988 No 159).

Schedules

[The full text of this Act contains the full text of these schedules. The full text is not reproduced in this book]
Schedule 1 Text of South Pacific Nuclear Free Zone Treaty of 6 August 1985
Schedule 2 Text of Treaty Banning Nuclear Weapon Tests in the Atmosphere, in Outer Space and Under Water of 5 August 1963
Schedule 3 Text of Treaty on the Non-Proliferation of Nuclear Weapons of 1 July 1968
Schedule 4 Text of Treaty on the Prohibition of the Emplacement of Nuclear Weapons and Other Weapons of Mass Destruction on the Sea-Bed and the Ocean Floor and in the Subsoil Thereof of 11 February 1971
Schedule 5 Text of Convention on the Prohibition of the Development, Production and Stockpiling of Bacteriological (Biological) and Toxin Weapons and on their Destruction of 10 April 1972
[The main text of this Act is as at 1st July 2013]

North Atlantic Council Statement on the Treaty on the Prohibition of Nuclear Weapons

D.1 Introduction

Shortly after TPNW was passed by UNGA and signed by a majority of countries in the world, NATO's North Atlantic Council issued (NATO, 2017). The statement is shown in the next section.a Statement on the Treaty on the Prohibition of Nuclear Weapons

D.2 Statement

At Warsaw in July 2016, the Alliance set out clear positions on the issues of nuclear deterrence and nuclear disarmament:

> "Allies emphasise their strong commitment to full implementation of the Nuclear Non-Proliferation Treaty (NPT). The Alliance reaffirms its resolve to seek a safer world for all and to create the conditions for a world without nuclear weapons in full accordance with all provisions of the NPT, including Article VI, in a step-by-step and verifiable way that promotes international stability, and is based on the principle of undiminished security for all. Allies reiterate their commitment to progress towards the goals and objectives of the NPT in its mutually reinforcing three pillars: nuclear disarmament, non-proliferation, and the peaceful uses of nuclear energy."

Regarding the prevailing international security environment, they further noted that:

> "After the end of the Cold War, NATO dramatically reduced the number of nuclear weapons stationed in Europe and its reliance on nuclear weapons in NATO strategy. We remain committed to contribute to creating the conditions for further reductions in the future on the basis of reciprocity, recognising that progress on arms control and disarmament must take into account the prevailing international security environment. We regret that the conditions for achieving disarmament are not favourable today."

Seeking to ban nuclear weapons through a treaty that will not engage any state actually possessing nuclear weapons will not be effective, will not reduce nuclear arsenals, and will neither enhance any country's security, nor international peace and stability. Indeed it risks doing the opposite by creating divisions and divergences at a time when a unified approach to proliferation and security threats is required more than ever.

The ban treaty is at odds with the existing non-proliferation and disarmament architecture. This risks undermining the NPT, which has been at the heart of global non-proliferation and disarmament efforts for almost 50 years, and the IAEA Safeguards regime which supports it. The crisis caused by North Korea underlines the importance of preserving and enhancing the existing framework of the NPT.

The ban treaty, in our view, disregards the realities of the increasingly challenging international security environment. At a time when the world needs to remain united in the face of growing threats, in particular the grave threat posed by North Korea's nuclear programme, the treaty fails to take into account these urgent security challenges.

The fundamental purpose of NATO's nuclear capability is to preserve peace, prevent coercion, and deter aggression. Allies' goal is to bolster deterrence as a core element of our collective defence and to contribute to the indivisible security of the Alliance. As long as nuclear weapons exist, NATO will remain a nuclear alliance.

We call on our partners and all countries who are considering supporting this treaty to seriously reflect on its implications for international peace and security, including on the NPT.

As Allies committed to advancing security through deterrence, defence, disarmament, non-proliferation and arms control, we, the Allied nations, cannot support this treaty. Therefore, there will be no change in the legal obligations on our countries with respect to nuclear weapons. Thus we would not accept any argument that this treaty reflects or in any way contributes to the development of customary international law.

France, UK, US Statement After UNGA Adoption of TPNW

E.1 Introduction

Following the adoption of the TPNW by UNGA, the UN representatives of France, UK, and USA issued a joint statement (P3, 2017) about the TPNW, as set out in the following Section.

E.2 Joint Press Statement from the Permanent Representatives to the United Nations of the United States, United Kingdom, and France Following the Adoption

New York City
July 7, 2017
FOR IMMEDIATE RELEASE

France, the United Kingdom and the United States have not taken part in the negotiation of the treaty on the prohibition of nuclear weapons. We do not intend to sign, ratify or ever become party to it. Therefore, there will be no change in the legal obligations on our countries with respect to nuclear weapons. For example, we would not accept any claim that this treaty reflects or in any way contributes to the development of customary international law. Importantly, other states possessing nuclear weapons and almost all other states relying on nuclear deterrence have also not taken part in the negotiations.

This initiative clearly disregards the realities of the international security environment. Accession to the ban treaty is incompatible with the policy of nuclear deterrence, which has been essential to keeping the peace in Europe and North Asia for over 70 years. A purported ban on nuclear weapons that does not address the security concerns that continue to make nuclear deterrence necessary cannot result in the elimination of a single nuclear weapon and will not enhance any country's security, nor international peace and security. It will do the exact opposite by creating even more divisions at a time when the world needs to remain united in the face of growing threats, including those from the DPRK's ongoing proliferation efforts. This treaty offers no solution to the grave threat posed by North Korea's nuclear program, nor does it address other security challenges that make nuclear deterrence necessary. A ban treaty also risks undermining the existing international security architecture which contributes to the maintenance of international peace and security.

We reiterate in this regard our continued commitment to the Treaty on the Non-Proliferation of Nuclear Weapons (NPT) and reaffirm our determination to safeguard and further promote its authority, universality and effectiveness. Working towards the shared goal of nuclear disarmament and general and complete disarmament must be done in a way that promotes international peace and security, and strategic stability, based on the principle of increased and undiminished security for all.

We all share a common responsibility to protect and strengthen our collective security system in order to further promote international peace, stability and security.

* * * * *

P5 Statement on TPNW

F.1 Introduction

In response to TPNW, the permanent 5 members of the UN Security Council released a joint statement reaffirming their commitment to the Treaty on the Non-Proliferation of Nuclear Weapons (P5, 2018), as set out in the following Section.

F.2 P5 Joint Statement on the Treaty on the Non-Proliferation of Nuclear Weapons

24 October 2018

We, the nuclear weapon States recognized by the Treaty on the Non-Proliferation of Nuclear Weapons, reaffirm our commitment to the Treaty, in all its aspects, fifty years since its signature.

This landmark Treaty has provided the essential foundation for international efforts to stem the threat that nuclear weapons would spread across the globe, and has thereby limited the risk of nuclear war. It has provided the framework within which the peaceful uses of nuclear technology – for electricity, medicine, agriculture and industry – could be promoted and shared, to the benefit of humanity. And by helping to ease international tensions and create conditions of stability, security and trust among nations, it has allowed for a vital and continuing contribution to nuclear disarmament.

We pledge our full and continued support for the work of the International Atomic Energy Agency (IAEA), which plays a critical role in NPT implementation, both in promoting the fullest possible cooperation on the peaceful uses of nuclear technology and in applying safeguards and verifying that nuclear programmes are exclusively for peaceful purposes. We emphasise the need to further strengthen the IAEA safeguards system, including the universalisation of the Additional Protocol.

We remain committed under the Treaty to the pursuit of good faith negotiations on effective measures related to nuclear disarmament, and on a treaty on general and complete disarmament under strict and effective international control. We support the ultimate goal of a world without nuclear weapons with undiminished security for all. We are committed to working to make the international environment more conducive to further progress on nuclear disarmament.

It is in this context that we reiterate our opposition to the Treaty on the Prohibition of Nuclear Weapons. We firmly believe that the best way to achieve a world without nuclear weapons is through a gradual process that takes into account the international security environment. This proven approach to nuclear disarmament has produced tangible results, including deep reductions in the global stockpiles of nuclear weapons.

The TPNW fails to address the key issues that must be overcome to achieve lasting global nuclear disarmament. It contradicts, and risks undermining, the NPT. It ignores the international security context and regional challenges, and does nothing to increase trust and transparency between States. It will not result in the elimination of a single weapon. It fails to meet the highest standards of non-proliferation. It is creating divisions across the international non-proliferation and disarmament machinery, which could make further progress on disarmament even more difficult.

We will not support, sign or ratify this Treaty. The TPNW will not be binding on our countries, and we do not accept any claim that it contributes to the development of customary international law; nor does it set any new standards or norms. We call on all countries that are considering supporting the TPNW to reflect seriously on its implications for international peace and security.

Rather, we urge all States to commit to the continued success of the NPT: to ensure compliance, to promote universalization, to ensure the highest standards of non-proliferation, and to respond to ongoing and emerging proliferation challenges, wherever they occur. In this context our five countries reiterate our commitment to continue our individual and collective efforts within the NPT framework to advance nuclear disarmament goals and objectives.

* * * * *

References

Commentary

Preparing this list of references was particularly problematic because of the custom for writers of legal materials to use footnote referencing, and for writers in subjects like social sciences, business, and management, to use author-date referencing. Further references can be found in the sister book NWIL3. For a bibliography about nuclear weapons see O'Donnell (2020).

* * * * *

References

Alperovitz; Gar (1995) *The Decision to Use the Atomic Bomb and the Architecture of an American Myth.* Mew York: Alfred A. Knopf Inc

Archer, Colin (2020) "A short history of the International Peace Bureau and its role in promoting international law" *in* Darnton et al, 2020 (Chapter 10)

Bennion, Francis A R. (2008) *Bennion on Statutory Interpretation.* London: Butterworths Law.

Boyle, Francis A (1986). "Relevance of International Law to the Paradox of Nuclear Deterrence." Northwestern University Law Review, vol. 80(5), 1985-1986, pp 1407-1448.

Boyle, Francis A. (2002) *The Criminality of Nuclear Deterrence.* Atlanta, GA: Clarity Press.

Burroughs, John (1997) *The Legality of Threat or Use of Nuclear Weapons: A Guide to the Historic Opinion of the International Court of Justice*, Munster: LIT Verlag.

BWC (1972) Convention on the Prohibition of the Development, Production and Stockpiling of Bacteriological (Biological) and Toxin Weapons and on their Destruction. New York: United Nations.

Charles River Editors (2019) *The Soviet Nuclear Weapons Program: The History and Legacy of the USSR's Efforts to Build the Atomic Bomb*, Charles Rivers Editors.

Chinkin, Chritine and Arimatsu, Louise (2021) *Legality Under International Law of the United Kingdom's Nuclear Policy As Set Out*

In The 2021 Integrated Review: Joint Opinion. London: Campaign for Nuclear Disarmament.

Clark, Ronald W. (1961) *The Birth of the Bomb: The Untold Story of Britain's Part in the Weapon that Changed the World.* London: Scientific Book Club.

CWC (1993) T*he Convention on the Prohibition of the Development, Production, Stockpiling and Use of Chemical Weapons and on their Destruction* (Chemical Weapons Convention). New York: United Nations.

Darnton, Geoffrey (ed.) (1989) *The Bomb and the Law: London Nuclear Warfare Tribunal: Evidence, Commentary and Judgment,* Stockholm: Alva and Gunnar Myrdal Foundation & Swedish Lawyers Against Nuclear Arms.

Darnton, Geoffrey (2005) "Content Analysis as a Tool of Information Warfare". *Journal of Information Warfare* 4(2) pp 1- 11

Darnton, Geoffrey; Archer, Colin; Falk, Richard A; Grief, Nick; Krieger, David; McBride, Seán (2020) *Nuclear Weapons and International Law.* (3rd ed.) Bournemouth, UK: Durotriges Press.

Datan, Merav and Ware, Alyn (1999) Securing our Survival: The Case for a Nuclear Weapons Convention The Model Convention on the Prohibition of the Development, Testing, Production, Stockpiling, Transfer, Use and Threat of Use of Nuclear Weapons and on their Elimination. Cambridge, MA: IPPNW (with IALANA and INESAP).

Datan, Merav; Hill, Felicity; Scheffran, Jürgen and Ware, Alyn (2007) Securing our Survival (SOS): The Case for a Nuclear Weapons Convention Including the Updated Model Convention on the Prohibition of the Development, Testing, Production, Stockpiling, Transfer, Use and Threat of Use of Nuclear Weapons and on their Elimination. Cambridge, MA: IPPNW (with IALANA and INESAP).

Davis, Nicola (2017) "Soviet submarine officer who averted nuclear war honoured with prize". Manchester: The Guardian 27th October 2017. Available at: https://www.theguardian.com/science/2017/oct/27/vasili-arkhipov-soviet-submarine-captain-who-averted-nuclear-war-awarded-future-of-life-prize .

Dewes, C.F. (1998) *The World Court Project: the evolution and impact of an effective citizen's movement,* PhD Dissertation. Armidale, Australia: University of New England. Available at: https://e-publications.une.edu.au/vital/access/manager/Repository/une:17024. [May 2015].

Drummond, Brian (2019) "UK Nuclear deterrence policy: an unlawful threat of force". *Journal on the Use of Force and International Law,* Vol 6(2), pp193-241, DOI:10.1080/20531702.2019.1669323

Duling, Kaitlyn (2019) *The Order to Drop the Atomic Bomb, 1945.* New York: Cavendish Square Publishing.

Falk, Richard A. (1965) "The Shimoda Case: A Legal Appraisal of the Atomic Attacks Upon Hiroshima and Nagasaki", *The American Journal of International Law*, Vol. 59, No. 4, pp. 759-793.

Falk, Richard A. (1971) "Beyond Deterrence: The Quest for World Peace" *in* Falk, Richard A., *This Endangered Planet: Prospects and Proposals for Human Survival*, New York: Random House.

Falk, Richard A. (1983) "Toward a Legal Regime for Nuclear Weapons". *McGill Law Journal* v28 pp 519-541.

Falk, Richard A (2017) "Challenging Nuclearism: The Nuclear Ban Treaty Assessed ". *Asia-Pacific Journal*, v15(14)1.

Farebrother,, George (ed.) (2012) *Freedom from Nuclear Weapons Through Legal Accountability and Good Faith Conference Report*. Nottingham: Spokesman for the World Court Project UK

Forsyth, Robert (2020) *Why Trident?* Nottingham: Spokesman Books.

From Scratch (1993) *Pacific 3.2.1. Zero* (musical film). Available via Vimeo at: https://vimeo.com/29293660. [Accessed March 2121].

Gale (2001) *A Study Guide for Richard Rhodes's "The Making of the Atomic Bomb"*. Farmington Hills, MI: Creative Media Partners in partnership with Gale Cengage Learning.

Gallagher, Thomas (2002) *Assault In Norway: Sabotaging the Nazi Nuclear Program*, Lyons Press

Gilling, Tom and McKnight, John (1991) *Trial and Error: Mordechai Vanunu and Israel's Nuclear Bomb*. Monarch Books.

Green, Robert (2018) *Security Without Nuclear Deterrence* (2nd ed.), Nottingham: Spokesman Books.

Grotius, Hugo (1682) *Treating of the Rights of War & Peace. In the first is handled, Whether any War be Just. In the second is shewed, The Causes of War, both Just and Unjust. In the Third is declared, What in War is Lawful; that is, Unpunishable*. London: Thomas Basset and Ralph Smith. [trans. William Evats].

Hajnoczi, Thomas (2020) "The Relationship between the NPT and the TPNW". Journal for Peace and Nuclear Disarmament 3(1), pp 87-91.

Hoon, Geoffrey(2002) "UK 'prepared to use nuclear weapons'". London: BBC News 20th March 2002. Available at: http://news.bbc.co.uk/1/hi/uk_politics/1883258.stm. *See also* "Select Committee on Defence Minutes of Evidence Examination of Witnesses (Questions 220-238)". London: UK Parliament Available at: https://publications.parliament.uk/pa/cm200102/cmselect/cmdfence/644/2032008.htm.

Hudson,Kate (2018) *CND at 60:Britain's Most Enduring Mass Movement*. London: Public Reading Rooms.

Housmans (2021) *Housmans Peace Diary with World Peace Directory* (68th ed. 2021). London: Housmans Bookshop. Updated annually.

ICC (1998) *Rome Statute of the International Criminal Court*, The

178

Hague: International Criminal Court.

ICC (2011) *Rome Statute of the International Criminal Court,* Den Haag: International Criminal Court.. Updated. Available online at: https://www.icc-cpi.int/resource-library/documents/rs-eng.pdf.

ICRC (2005a) Customary International Humanitarian Law Volume I: Rules. Geneva: International Committee of the Red Cross (ICRC) and, Cambridge: Cambridge University Press.

ICRC (2005b) Customary International Humanitarian Law Volume II: Practice. Geneva: International Committee of the Red Cross (ICRC) and, Cambridge: Cambridge University Press.

ILC (1950a) Yearbook of the International Law Commission 1950, Vol. II, New York: United Nations. [not actually published until 1957]. Available at: https://legal.un.org/ilc/publications/yearbooks/english/ilc_1950_v2.pdf

ILC (1950b) "Formulation Of Nurnberg Principles" in ILC (1950a) pp 181 - 195. Also available in summary form as: *Principles of International Law Recognized in the Charter of the Nürnberg Tribunal and in the Judgment of the Tribunal,* New York: United Nations 2005 printing available at: https://legal.un.org/ilc/texts/instruments/english/draft_articles/7_1_1950.pdf.

ILC (1950c) "Draft Code of Offences against The Peace and Security of Mankind" in ILC (1950a) pp 249-362.

ILC (1950d) Report of the International Law Commission to the General Assembly in ILC (1950a) pp 374 - 378

ILC (1996) Yearbook of the International Law Commission 1996, Vol. II pt 2, New York: United Nations. Available at: https://legal.un.org/ilc/publications/yearbooks/english/ilc_1996_v2_p2.pdf.

Jabber, Fuad (1971) *Israel and Nuclear Weapons.* London: Chatto & Windus for Institute of Strategic Studies.

Johnson, Rebecca and Zelter, Angie (eds.) (2011) *Trident and International Law: Scotland's Obligations,* Edinburgh: Luath Press Ltd.

Kahan, D. M. (1999). The secret ambition of deterrence. Harvard Law Review, 113(2), pp 413-500

Kelly, Cynthia C. (ed.) (2009) *The Manhattan Project: The Birth of the Atomic Bombin the Words of Its Creators, Eyewitnesses,and Historians.* New York: Black Dog & LeventhalPublishers.

Krieger, David. (2013) *Zero: The Case for Nuclear Weapons Abolition.* Santa Barbara, CA: Nuclear Age Peace Foundation, 2013.

Kuo, Kuang-Ming; Talley, Paul C; Huang, Chi-Hsien (2020) "A meta-analysis of the deterrence theory in security-compliant and security-risk behaviors", *Computers & Security,*Volume 96, September 2020. Elsevier ScienceDirect.

Lippman, Matthew (1986) "Nuclear Weapons and International Law: Towards a Declaration on the Prevention and Punishment of the

Crime of Nuclear Humancide". *Loyola of Los Angeles International and Comparative Law Review*, Vol 8(2) pp 183-234.

MacBride, Seán (1974) Nobel Lecture: The Imperatives of Survival. Stockholm: Nobel Foundation. Available at: https://www.nobelprize.org/prizes/peace/1974/macbride/lecture/.

Mills, Claire (2021) *Integrated Review 2021: Increasing the cap on the UK's nuclear stockpile.* London: House of Commons Library Briefing Paper 9175 19th March 2921.

Moruroa (2021) *Muroroa Files: Investigation into French Nuclear Tests in the Pacific.* INTERPRT, DISCLOSE, and Program on Science and Global Security at Princeton University. Available at: https://moruroa-files.org/en/investigation/moruroa-files. [Accessed March 2021].

Moscow (1943) "Moscow Declaration on Atrocities", part of Declaration of The Four Nations on General Security *in* Uniter Nations Documents 1941-1945, London: Royal Institute for International Affairs.

NATO (2017) North Atlantic Council Statement on the Treaty on the Prohibition of Nuclear Weapons. Available at: https://www.nato.int/cps/en/natohq/news_146954.htm [Accessed March 2021].

NTC (2005) *International Convention on the Suppression of Acts of Nuclear Terrorism.* New York: Uniter Nations. Available at: https://treaties.un.org/doc/db/terrorism/english-18-15.pdf.

Nuremberg (1945) *Agreement for the Prosecution and Punishment of the Major War Criminals of the European Axis, and Charter of the International Military Tribunal. London, 8 August 1945*, London: UK Foreign and Commonwealth Office.

Nuremberg (1946) *Judgment of the International Military Tribunal for the Trial of German Major War Criminals (with the dissenting opinion of the Soviet member), Nuremberg 30th September and 1st October 1946*, London: His Majesty's Stationery Office (Document Miscellaneous No. 12 (1946)).

NWC (2007) Convention on the Prohibition of the Development, Testing, Production, Stockpiling, Transfer, Use and Threat of Use of Nuclear Weapons and on Their Elimination. New Your: United Nations General Assembly document A/62/650.

NWIL3 *see* Darnton et al. 2020.

P3 (2017) *Joint Press Statement from the Permanent Representatives to the United Nations of the United States, United Kingdom, and France Following the Adoption.* Available at: https://usun.usmission.gov/joint-press-statement-from-the-permanent-representatives-to-the-united-nations-of-the-united-states-united-kingdom-and-france-following-the-adoption/. [Accessed March 2021].

P5 (2018) *P5 Joint Statement on the Treaty on the Non-Proliferation of*

180

Nuclear Weapons. Available at: https://www.gov.uk/government/news/p5-joint-statement-on-the-treaty-on-the-non-proliferation-of-nuclear-weapons [Accessed March 2021].

Pal, Radhabinod (1048) "Judgment of the Honourable Mr. Justice Pal, Member for India" *in* Boister, Neil and Cryer, Robert (eds.) (2008) *Documents on the Tokyo International Military Tribunal: Charter, Indictment, and Judgments*, Oxford: Oxford University Press, pp 809-1426.

PICAT (2021) Public Interest Case Against Trident - website. Available at: https://picat.online/

Pictet, Jean S. (ed.) (1952) *I Geneva Convention for the Amelioration of the Conditions of the Wounded, and Sick in Armed Forces in the Field* (with commentary), Geneva: International Committee of the Red Cross.

Pictet, Jean S. (ed.) (1958) *IV Geneva Convention Relative to the Protection of Civilian Persons in Time of War,* (with commentary), Geneva: International Committee of the Red Cross.

Pictet, Jean S. (ed.) (1960a) *II Geneva Convention for the Amelioration of the Conditions of the Wounded ,Sick and Shipwrecked Members of Armed Forces at Sea* (with commentary) Geneva: International Committee of the Red Cross.

Pictet, Jean S. (ed.) (1960b) *III Geneva Convention Relative to the Treatment of Prisoners of War,* (with commentary), Geneva: International Committee of the Red Cross.

Popp, Roland; Horovitz, Liviu; Wenger, Andreas (eds.) (2017) *Negotiating the Nuclear Non-Proliferation Treaty: Origins of the Nuclear Order.* Abingdon, UK: Routledge.

Porter, David (2010) *Hitler's Secret Weapons 1933-1945:TheEssential Facts and Figures for Germany's SecretWeapons Programme. Chapter 8, pp 168-177.* London: Amber Books Ltd.

Raratonga (1985) South Pacific Nuclear Free Zone Treaty, Suve, Fiji: Pacific Islands Forum Secretariat. Available at: http://www.forumsec.org/wp-content/uploads/2018/02/South-Pacific-Nuclear-Zone-Treaty-Raratonga-Treaty-1.pdf.

Rhodes, Richard (1996) *The Making of the Atomic Bomb*, Simon and Schuster.

Richardson, Lewis Fry (1960) *Statistics of Deadly Quarrels* London: Stevens & Sons Ltd.

RIFE (2019) Radioactivity in Food and the Environment, 2018. Multiple Agencies: Environment Agency, Food Standards Agency, Food Standards Scotland, Natural Resources Wales, Northern Ireland Environment Agency, Scottish Environment Protection Agency. Available at: https://www.foodstandards.gov.scot/downloads/RIFE_24.pdf [accessed March 2021].

Rosenne, Shabtai (1967) "The Depositary of International Treaties".

The American Journal of International Law, Vol. 61(4)4 pp. 923-945. Cambridge: Cambridge University Press.

Savranskaya, Svetlana V. (2005) "New Sources on the Role of Soviet Submarines in the Cuban Missile Crisis". *Journal of Strategic Studies* 28(2) pp 233-259. Routledge.

Shimoda (1963) "Ryuichi Shimoda et al. v. The State - Tpkyo District Cout decision 7th December, 1963" translation in *The Japanese Annual of International Law No 8* 1964, pp212-252. Tokyo: The Japan Branch of the International Law Association.

Stewart, Luke J. (2014) *"A New Kind of War": The Vietnam War and The Nuremberg Principles, 1964-1968*. PhD thesis. Waterloo, Ontario, Canada: University of Waterloo.

Stone, Jon (2021) "Boris Johnson 'violating international law' with plan to build more nuclear weapons: Defence review appears to breach Article 6 of nuclear non-proliferation treaty". London: Independent, 16th March 2021. Available at: https://www.independent.co.uk/news/uk/politics/boris-johnson-uk-nuclear-weapons-international-law-b1817827.html.

Szasz, Feren Morton (1992) *British Scientists and the Manhattan Project: The Los Alamos Years*, Basingstoke: Macmillan.

UK Govt (2021) *Global Britain in a competitive age: The Integrated Review of Security, Defence, Development and Foreign Policy*. London: UK Government.

UNGA/95 (1946) "Affirmation of the Principles of International Law recognized by the Charter of the Niirnberg Tribunal", New York: United Nations General Assembly. Available online at: https://crimeofaggression.info/documents//6/1946_GA_Resolution_95.pdf

UNGA/177, (1947) "Formulation of the principles recognized in the Charter of the Niirnberg Tribunal and in the judgment of the Tribunal",New York: United Nations General Assembly. Available online at: https://legal.un.org/docs/?symbol=A/RES/177(II)

UNGA/3314 (1974) "Definition of Aggression" New York: United Nations General Assembly.

US-UK (1958) Agreement between the Government of the United Kingdom of Great Britain and Northern Ireland and the Government of the United States of America for Co-operation on the Uses of Atomic Energy for Mutual Defence Purposes. Treaty Series No 41 (1958). London: HM Stationery Office. Available at: https://www.cvce.eu/content/publication/2014/6/12/a1ee4c1f-2166-48f3-a886-2711bd647111/publishable_en.pdf [this document is updated from time to time].

Venter, Al J. (2008) *How South Africa built six atom bombs and then abandoned its Nuclear Weapons program*. Ashanti Publishing Group.

VCLT (1969) Vienna Convention on the Law of Treaties. New York: United Nations. Entered into force 27th January, 1980.

Villa, Brian L. and Bonnett, John (1996) "Understanding Indignation:

Gar Alperovitz, Robert Maddox, and the Decision to Use the Atomic Bomb". In Reviews in American History, , 24(3) pp. 529-536. Baltimore, MD: Johns Hopkins University Press.

Wikipedia contributors. (2021, February 15). "South Africa and weapons of mass destruction". *In* Wikipedia, The Free Encyclopedia. Retrieved 18:53, February 23, 2021, from https://en.wikipedia.org/w/index.php?title=South_Africa_and_weapons_of_mass_destruction&oldid=1006907385

Wilcox, Robert K. (2019) *Japan's Secret War: How Japan's Race to Build Its Own Atomic Bomb Provided the Groundwork for North Korea's Nuclear Program* (3rd ed.). Brentwood, TN: Permuted Press;

WNA (2020) "Outline History of Nuclear Energy", London: World Nuclear Association. Available at: https://www.world-nuclear.org/information-library/current-and-future-generation/outline-history-of-nuclear-energy.aspx. Accessed February 2021 (this article is updated from time to time).

Wright, Quincy (1965) *A Study of War : Second Edition with a Commentary on War since 1942*, Chicago: University of Chicago Press. {for relevance to nuclear weapons, it is the Commentary, pp 1499-1577 of particular relevance.]

Zelter, Angie (2001) *Trident on Trial: the case for people's disarmament*, Edinburgh: Luath Press Ltd.

Index

M

MacBride, Seán 16, 176, 179
  Nobel Peace Prize address 16
MAD ix
Manhattan Project 7, 63
  UK contribution 63
Manifesto to Move Forward 77–80
Martens, de 30, 32
McKnight, John 15, 177
MDA ix
medical services 38
MID ix
Miller, Sue viii
Mills, Claire 64, 179
Montebello Islands
  first UK nuclear test 11
  UK nuclear test 65
Moon Treaty 1979 14
morality 24, 27
  law and 24
Moruroa 39, 179
Moscow Declaration 7, 51, 179
Move the Nuclear Weapons Money 37
mutual assistance
  between TPNW states parties 37
Mutual Defence Agreement
  US-UK 64

N

Nagasaki 26
NAPF ix, 12. *See* Nuclear Age Peace Foundation; *See also* Nuclear Age Peace Foundation
Nation States 21, 80
NATO ix, 32, 39, 179
  position re TPNW 4, 169–170
Necessity
  Principle of 6, 11, 28
negotiations
  to be in good faith 32
Neutrality
  Principle of 6, 13, 14, 16, 17, 28
New Zealand 2
  nuclear weapons free zone 4
New Zealand Nuclear Free Zone 14, 47–48
New Zealand Nuclear Free Zone, Disarmament, and Arms Control Act 1987.

Article 5 Prohibitions 48
  prohibitions compared with TPNW 48
Nick Grief. *See* Grief, Nicholas
No Ecocide
  Principle of 6, 14, 16, 17, 19, 20, 21, 22, 28
Non-Proliferation Treaty, 1968 1, 32
  Article VI 13
  does not permit NWS 34
North Atlantic Council Statement on the Treaty on the Prohibition of Nuclear Weapons 169–170
Norway
  heavy water production 6
NPT ix
  Depositary Governments, and 76
  limited to nuclear weapons and explosive devices 64
  problems with 2
  vs TPNW 2, 40–44
NTC 179
  applies to individuels 44
  exemption of NWS probably invalid 44
Nuclear Age Peace Foundation 12
nuclear deterrence
  unlawful 33
nuclear power
  nuclear weapons, and 64
nuclear propulsion 36
nuclear reactors 36
Nuclear Reactors
  change the narrative 80
Nuclear Terrorism Convention. *See* NTC
Nuclear Warhead Capability Sustainment Programme 71
nuclear waste 38, 80
nuclear weapon
  delivery vehicles. 37
Nuclear Weapon Free Zones
  1967 Outer Space Treaty 14
  1967 Tlatelolco, Treaty of 14
  1971 Seabed Treaty 14
  1979 Moon Treaty 14
  1985 Rarotonga Treaty 14
  1987 New Zealand Nuclear Free Zone 47
  1995 Bangkok Treaty 14
  1995 Pelindaba Treaty 14

V

Vanguard submarines 67
van Riet, Rob viii
V bombers 65
VCLT x, 76, 181
Venter, Al J. 16, 181
victim assistance 36
Vienna Convention on the Law of
Treaties 76
Villa, Brian L. 8, 181
VSEL 66

W

war crimes 30, 52
Nuremberg Principles definition 33
Ware, Alyn viii, 35, 79, 176
WCP x, 1
launched in Geneva in May 1992 16
number of organizations involved 17
WE177 air-launched free-fall nuclear
bomb 66
Wenger, Andreas 13, 180
Wikipedia
on South Africa nuclear weapons 16,
181
Wilcox, Robert K. 15, 182
WILPF x
WNA x, 182. *See* World Nuclear As-
sociation
Women's International League for
Peace and Freedom 12, 13
World Court Project 12, 16–22. *See
also* WCP
World Nuclear Association 6, 182
World War III
prevented by nuclear weapons? 24
Wright, Quincy 27, 182
WWII x

Z

Zelter, Angie viii, 75, 178, 182

www.ingramcontent.com/pod-product-compliance
Lightning Source LLC
Chambersburg PA
CBHW060529210326
41519CB00014B/3183